Steroid Chemistry at a Glance

Other Titles Available in the *Chemistry at a Glance* series:

Chemical Thermodynamics at a Glance
H. Donald Brooke Jenkins
ISBN: 978-1-4051-3997-7

Natural Product Chemistry at a Glance
Stephen P. Stanforth
ISBN: 978-1-4051-4562-6

Heterocyclic Chemistry at a Glance
John A. Joule, Keith Mills
ISBN: 978-1-4051-3918-2

Environmental Chemistry at a Glance
Ian Pulford, Hugh Flowers
ISBN: 978-1-4051-3532-0

The Periodic Table at a Glance
Mike Beckett, Andy Platt
ISBN: 978-1-4051-3299-2

Organic Chemistry at a Glance
Laurence M. Harwood, John E. McKendrick, Roger Whitehead
ISBN: 978-0-86542-782-2

Chemical Calculations at a Glance
Paul Yates
ISBN: 978-1-4051-1871-2

Stereochemistry at a Glance
Jason Eames, Josephine M Peach
ISBN: 978-0-632-05375-9

Reaction Mechanisms at a Glance: A Stepwise Approach to Problem-Solving in Organic Chemistry
Mark G. Moloney
ISBN: 978-0-632-05002-4

Steroid Chemistry at a Glance

Daniel Lednicer

A John Wiley & Sons, Ltd., Publication

This edition first published 2011
© 2011 John Wiley & Sons Ltd

Registered office
John Wiley & Sons Ltd, The Atrium, Southern Gate, Chichester, West Sussex PO19 8SQ, United Kingdom

For details of our global editorial offices, for customer services and for information about how to apply for permission to reuse the copyright material in this book please see our website at www.wiley.com.

The right of the author to be identified as the author of this work has been asserted in accordance with the Copyright, Designs and Patents Act 1988.

All rights reserved. No part of this publication may be reproduced, stored in a retrieval system, or transmitted, in any form or by any means, electronic, mechanical, photocopying, recording or otherwise, except as permitted by the UK Copyright, Designs and Patents Act 1988, without the prior permission of the publisher.

Wiley also publishes its books in a variety of electronic formats. Some content that appears in print may not be available in electronic books.

Designations used by companies to distinguish their products are often claimed as trademarks. All brand names and product names used in this book are trade names, service marks, trademarks or registered trademarks of their respective owners. The publisher is not associated with any product or vendor mentioned in this book. This publication is designed to provide accurate and authoritative information in regard to the subject matter covered. It is sold on the understanding that the publisher is not engaged in rendering professional services. If professional advice or other expert assistance is required, the services of a competent professional should be sought.

The publisher and the author make no representations or warranties with respect to the accuracy or completeness of the contents of this work and specifically disclaim all warranties, including without limitation any implied warranties of fitness for a particular purpose. This work is sold with the understanding that the publisher is not engaged in rendering professional services. The advice and strategies contained herein may not be suitable for every situation. In view of ongoing research, equipment modifications, changes in governmental regulations, and the constant flow of information relating to the use of experimental reagents, equipment, and devices, the reader is urged to review and evaluate the information provided in the package insert or instructions for each chemical, piece of equipment, reagent, or device for, among other things, any changes in the instructions or indication of usage and for added warnings and precautions. The fact that an organization or Website is referred to in this work as a citation and/or a potential source of further information does not mean that the author or the publisher endorses the information the organization or Website may provide or recommendations it may make. Further, readers should be aware that Internet Websites listed in this work may have changed or disappeared between when this work was written and when it is read. No warranty may be created or extended by any promotional statements for this work. Neither the publisher nor the author shall be liable for any damages arising herefrom.

Library of Congress Cataloging-in-Publication Data

Lednicer, Daniel, 1929-
 Steroid chemistry at a glance / Daniel Lednicer.
   p. cm.
 Includes index.
 ISBN 978-0-470-66085-0 (cloth)
 1. Steroids. I. Title.
 QD426.L43 2010
 547'.7–dc22

                               2010025742

A catalogue record for this book is available from the British Library.

Print ISBN: 9780470660850 (hb), 9780470660843 (pb)
ePDF ISBN: 9780470973622
oBook ISBN: 9780470973639

Set in 10/12.5pt Times by Thomson Digital, Noida, India

Contents

Preface			**vii**
Introduction			**2**
1	**Steroids: a Brief History**		**10**
	1.1	Structure Determination	10
		1.1.1 Cholesterol and Cholic Acid	10
		1.1.2 The Sex Steroids	13
		1.1.3 Corticosteroids	16
2	**Sources of Steroids**		**20**
	2.1	Biosynthesis	20
	2.2	Commercial Steroid Starting Materials	22
		2.2.1 Diosgenin	23
		2.2.2 Soybean Sterols	25
3	**Estranes: Steroids in Which Ring A is Aromatic**		**28**
	3.1	Biological Activity	28
	3.2	Sources of Estranes	28
		3.2.1 From Androstanes	28
		3.2.2 Estrogens by Total Synthesis	32
	3.3	Chemical Reactions of Estranes	37
		3.3.1 Aromatic A-ring Reactions	37
		3.3.2 Modifications on Ring B	40
		3.3.3 Modifications on Ring C	41
		3.3.4 Modifications on Ring D	42
	3.4.	Some Drugs Based on Estranes	45
4	**Gonanes or 19-nor-Steroids**		**48**
	4.1	Preparation of Gonane Starting Materials	48
		4.1.1 Birch Reduction	48
		4.1.2 Synthesis by Sequential Annulation Reactions	49
	4.2	Anabolic–Androgenic Gonanes	50
		4.2.1 Biological Activity	50
		4.2.2 Synthesis of 19-Norandrogens	51
	4.3	Progestational Gonanes	55
		4.3.1 Biological Activity	55
		4.3.2 Preparation of 19-Norprogestins	55
	4.4	Some Drugs Based on Gonanes	67
		4.4.1 Androgenic–Anabolic Agents	67
		4.4.2 Progestins	67
		4.4.3 Progestin Antagonists	67
5	**Androstanes, C_{19} Steroids and Their Derivatives**		**68**
	5.1	Biological Activity	68
	5.2	Sources of Androstanes	68
		5.2.1 From Pregnenolone	68
		5.2.2 Fermentations	69
		5.2.3 Total Synthesis	69

vi Contents

	5.3	Modified Anabolic–Androgenic Androstanes	69
		5.3.1 17-Desalkyl Compounds	69
		5.3.2 17-Alkyl Compounds	73
		5.3.3 Modifications on Ring B	77
		5.3.4 Modifications on Ring C	79
		5.3.5 Modifications on Ring D	80
	5.4	17-Spirobutyrolactone Aldosterone Antagonists	83
	5.5	Some Drugs Based on Androstanes	85
		5.5.1 Androgens	85
		5.5.2 Spirobutyrolactones	85

6 Pregnanes, Part 1: Progestins — **86**
- 6.1 Biological Activity — 86
- 6.2 Sources of Progesterone — 86
  - 6.2.1 From Phytochemicals — 86
  - 6.2.2 By Total Synthesis — 87
  - 6.2.3 From Dehydroepiandrosterone (DHEA) Acetate — 88
- 6.3 Modified Pregnanes — 88
  - 6.3.1 17-Hydroxy and Acyloxy Derivatives — 88
  - 6.3.2 Modifications on Ring A — 89
  - 6.3.3 Modifications on Ring B — 90
  - 6.3.4 General Methods for Modifications on Ring D — 94
  - 6.3.5 More Progesterone Analogues — 96
- 6.4 Some Drugs Based on Progestins — 100
  - 6.4.1 Medroxyprogesterone Acetate (10-2) — 100
  - 6.4.2 Megestrol Acetate (10-3) — 101
  - 6.4.3 Melengestrol Acetate (26-7) — 101

7 Pregnanes, Part 2: Corticosteroids — **102**
- 7.1 Biological Activity — 102
- 7.2 Sources of Corticoids — 102
  - 7.2.1. Introduction of Oxygen at C_{11} — 102
  - 7.2.2 Construction of the Dihydroxyacetone Side Chain — 103
- 7.3 Modified Corticoids — 105
  - 7.3.1 Unsaturation — 105
  - 7.3.2 Additional Alkyl Groups — 106
  - 7.3.3 Halogenated Corticoids — 109
  - 7.3.4 Hydroxylation: 16,17-Diols — 111
  - 7.3.5 Corticoids with Multiple Modifications — 112
  - 7.3.6 Miscellaneous Corticoids — 117
- 7.4 Some Drugs Based on Corticoids — 119

8 Miscellaneous Steroids — **122**
- 8.1 Heterocyclic Steroids — 122
  - 8.1.1 Introduction — 122
  - 8.1.2 Steroids with a Heteroatom in Ring A — 122
  - 8.1.3 Steroids with a Heteroatom in Ring B — 124
  - 8.1.4 Steroids with a Heteroatom in Ring C — 126
  - 8.1.5 Steroids with a Heteroatom in Ring D — 128
- 8.2 Cardenolides — 129
  - 8.2.1 Actodigin Aglycone — 130
  - 8.2.2 Synthesis from a Bile Acid — 130
- 8.3 Compounds Related to Cholesterol — 132

Subject Index — **135**
Reactions Index — **141**

Preface

The term 'steroid' has become virtually synonymous with androgenic–anabolic compounds (mainly analogues of testosterone) to the majority of the public. The sport sections of many newspapers carry almost daily exposés of the usage of these drugs by athletes seeking to enhance their performance. The androgens in question, however, comprise only a single, relatively small, class of biologically active steroids. What may be called the athletic androgens are in reality overshadowed by a large universe of compounds that share the same tetracyclic nucleus. The term 'androgen' in fact represents only one-tenth (1.4×10^6 versus 13.5×10^6) of the hits when Googling the term 'steroid'. The very sizeable number of steroids that are approved by regulatory agencies as therapeutic drugs eclipses the group of legal androgenic–anabolic drugs.

By the 1940s, about a decade after their structure had been firmly established, it became evident that steroids might well comprise a structural lead for drug design. Preliminary results from pharmacological studies, carried out at that time, suggested that selected steroids could potentially lead to drugs aimed at targets as diverse as oral contraceptives on the one hand and inflammation on the other. The potential markets for such drugs spurred major chemical efforts in industrial and to some extent in academic laboratories. Research that led to steroid-based therapeutic agents was carried out largely in the laboratories of 'Big Pharma' over the two decades following the end of World War 2. This resulted in the accretion of a large body of organic chemistry often denoted 'Steroid Chemistry', and also a sizable number of new therapeutic agents. The assignment of a USAN designation, more familiarly known as a generic name, to a potential drug indicates that the sponsor intends to take the initial steps to assess the clinical activity of the compound. Close to 130 steroids have been assigned official USAN non-proprietary names.

Reports of side-effects that accumulated as the drugs became more widely used led chemists to go back to modify the structures of the offending agent in the hope of producing better tolerated entities. It would be naïve to dismiss the aim of obtaining a place in the market by means of one's own proprietary and patented entity as additional motivation for that task. It became evident by the mid-1970s that many of the undesired properties, that is, side-effects, were often simply another aspect of the desired hormonal activity. Research aimed at novel steroid-based drugs consequently decreased markedly. The preceding chemical research in the area had by the accumulated a significant body of specialized reactions.

All steroids, be they derived from natural sources or produced by total synthesis, share the same rigid, fixed, three-dimensional framework. Many of the chemical properties of steroids, such as the dependence of the reactivity of functional groups on their specific location, are determined by steric properties of the steroid nucleus. That nucleus incorporates over half a dozen chiral centers not counting the side chain. Cholesterol, for example, can in theory consist of no fewer than 512 stereoisomers. This compound actually occurs as a single chiral species, as do virtually all other steroid-based products. The chemistry of these compounds thus provides a rich source for the study of the effects of stereochemistry on chemical reactivity. The reactivity of a pair of ketones in the same molecule, for example, will often differ markedly due to differences in their steric milieu. Structural features of steroids generally determine biological activity. Steroids with an aromatic A ring will, for example, act as estrogens. Differing structural features found in each of those groups has a marked influence on the reactions and reaction sequences used in preparing potential drugs. A text on steroid chemistry could in theory be organized either on the basis of reactions or alternatively by structural class. Grouping compounds by reaction-based sections it is felt would lead to somewhat jumbled presentations. Many of the organic reaction schemes in used steroid chemistry are characteristic of one or another of the broad structural classes. This volume is accordingly divided into the traditional broad structural chapters. The circumstance that biological activity follows the same organization merely illustrates the concordance of structure and activity.

Rules of nomenclature appear early on in many beginning organic chemistry texts. In somewhat the same vein, the Introduction to this book starts with the conventions for naming steroids. This is followed by a concise account of the molecular mechanism of action by which many steroids exert their biological effects. More detailed descriptions of the activity of these compounds will be found in the opening paragraphs of the individual structural classes.

Chapter 1 describes the history of steroids with particular attention to the research aimed defining the chemical structure of what were at the time fairly complex molecules. The reader may find it convenient to skim over this section at first reading and to then return after acquiring more familiarity with steroid chemistry.

Chapter 2 opens with a description of the biosynthesis of naturally occurring steroids. The conversion of two very different phytochemicals to steroids that can be elaborated to potential drugs follows. The narrative focuses on a discussion of the chemistry whereby these steroidal natural products are modified into steroid starting materials.

Specifically, this describes first the chemistry used to convert diosgenin from Mexican yam roots to dehydropregnenolone and then a discussion of the preparation of pregnenolone from stigmasterol obtained from soybeans.

Chemical manipulations of aromatic A-ring steroids, the estranes, are described in Chapter 3. The relative simplicity of the structure of estranes has led to the development of close to half a dozen syntheses that differ in approach, starting from laboratory chemicals. One of these total syntheses, in contrast to those found in subsequent chapters, is used in actual practice to prepare intermediates for the gonanes discussed in Chapter 4. There follows a description of the process long used to prepare aromatic A-ring steroids from phytochemical-derived sources. Chemical reactions of estranes close this chapter.

The chemistry of gonanes, more familiarly known as 19-nor steroids, constitutes the subject matter of Chapter 4. This chapter opens with a discussion the general methods used to prepare the gonane nucleus. Those methods include two syntheses starting from laboratory chemicals that differ markedly in their approach. The description of the chemistry of the gonanes is divided according the disparate biological activity structural variants. These comprise a section on compounds that act as androgenic–anabolic agents and another that includes progestational agents. This last section includes most of the oral contraceptives. A discussion of the newer 11-arylgonane progesterone antagonist concludes Chapter 4.

The androstanes, often called C-19 steroids in that they include methyl groups at each of the angular carbons at C_{10} and C_{13}, are described in Chapter 6. This section, like the preceding one, also starts with a description of the chemistry used to provide starting material. It also includes a discussion of a total synthesis based on an electrocyclic reaction. The bulk of the chapter comprises of a ring-by-ring description of the chemistry that has been used to prepare modified C-19 androstanes. The bulk of the compounds in this first part of the chapter exhibit androgenic-anabolic activity. Incorporation of a spirobutyrolactone at C_{17} of the C-19 androstanes leads to compounds that act as diuretics as a result of their aldosterone blocking activity. A brief section on those compounds completes Chapter 5.

Pregnanes in essence comprise androstanes that in addition feature a two-carbon side chain, almost exclusively acetyl, at C_{17}. This carbon skeleton is common to both progestins and corticosteroids. Chapter 6 is devoted to a discussion of the chemistry used to prepare derivatives of the simpler of the two, progesterone. Sources of starting materials for modified progestins from both diosgenin and stigmasterol are discussed at the beginning of the chapter. This is followed by a total synthesis that includes a cascade electrocyclization reaction that somewhat resembles the biosynthetic process by which squalane goes to lanosterol. There follows a ring-by-ring examination of the chemistry used to modify the basic pregnane nucleus.

Structurally more complex corticosteroids, commonly called corticoids, are grouped in Chapter 7, the second section on pregnanes. The biological activity of this class of pregnanes depends on the presence of an oxygen atom, either as a ketone or as an alcohol, at C_{11} in ring C. The rarity of this structural feature in Nature placed high priority on developing methods for adding that feature to more abundant steroids from Nature. Chapter 7, as in the preceding chapter, opens with a discussion of the methods that have been developed for preparing the starting 11-oxypregnanes required for both clinical supplies and research on analogues. Methods for preparing analogues that include single modifications are considered first. Corticoids comprise one of the rare classes of compounds in which the potentiating effects of structural changes are additive. The closing sections of Chapter 7 discuss the chemistry for preparing compounds with multiple modification.

Groups of steroids that are too small for a full chapter are to be found in the seemingly inevitable chapter termed Miscellaneous. The first section considers steroids in which one of the ring carbon atoms is replaced by a heteroatom, more specifically oxygen or nitrogen. Two compounds in this class, both androstanes, are approved for use in humans. Cardenolides, the steroid-based compounds obtained on removal of the sugars from the so-called cardiac glycosides, are considered in the next section. The chapter closes with a brief discussion of the chemistry involved in modifying the unsaturated cholestanes related to vitamin D.

Dan Lednicer
North Bethesda, MD

Introduction

The isoprene unit (Figure 1) is one of the ubiquitous naturally occurring hydrocarbons. This five carbon latent diene serves as starting material for a host of natural products in both plants and animals. The reactive diphosphate of this five carbon unit, often called isoprene pyrophosphate (IPP), readily couples with itself to form dimers that comprise a bewildering variety of open chain, cyclic, polycyclic and bridged polycyclic compounds known collectively as terpenes. The first coupling product, the straight chain C_{10} dimer may react further by either adding another isoprene or by condensing with other activated terpenes. Products from IPP range all the way at one extreme of a single isoprene to the polyisoprenoid, rubber, at the other extreme.

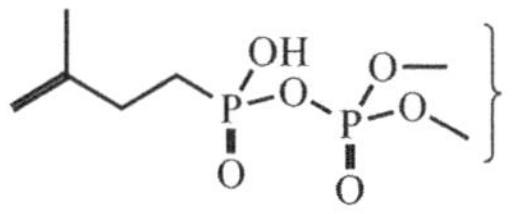

Figure 1 Isoprene unit (isopentenyl diphosphate)

Steroids comprise the singular set of effector molecules, crucial for life processes, derived from one of the multitude of other C_{30} triterpenes (see Chapter 2). These compounds, which arise from further transformations of the triterpene lanosterol, share a common rigid four-ring carbon skeleton. The association of steroids with vertebrates is believed to date back at least 540 million years, as shown by the detection of steroid-derived compounds, called steranes, in ancient fossils. The majority of the endogenous animal steroids are considered hormones since they control various bodily functions at very low concentrations. Estrogens, one of the first groups to be identified, are secreted mainly by the ovary. These compounds control reproductive function in females and also maintenance of female genitalia. Progesterone, secreted by the corpus luteum on that same organ, is involved in many of the same functions as the estrogens and in addition supports pregnancy. The male counterpart to those compounds, testosterone, secreted mainly by the testes, controls the production of sperm and maintains male genitalia; the anabolic, nitrogen-conserving activity of androgens enhances muscle mass. The structurally more complex glucocorticoids, such as cortisone, which are secreted by the adrenal cortex, act on glucose metabolism and to some extent mineral levels in blood. Aldosterone, structurally yet more complex, acts directly on the kidney to maintain electrolytes and blood volume. Ergosterol, which is converted to vitamin D by exposure to sunlight, controls calcium levels and consequently bone health.

Cholesterol, whether ingested or formed endogenously, provides the starting material for the biosynthesis of all the other steroids found in mammals. Excess cholesterol is oxidized in the liver to polar compounds called cholic acids. This process converts the terminal side chain in cholesterol to a carboxylic acid and introduces hydroxyl groups. These polyhydroxylated, steroidal acids play a central role in absorption of fats from the intestine and also excretion of superfluous cholesterol.

Nomenclature

Structural Bases for Naming Steroids

Steroids are designated by names that range from those based on IUPAC rules to the trivial (the term 'trivial' as used in the following paragraphs is not a value judgment but instead notes that the relevant name does not conform to formal naming conventions). The rigorous IUPAC name sidesteps the steroid nucleus entirely, naming these compounds as derivatives of cyclopenta[*a*]phenanthrene; the IUPAC name for estrone, for example, is (8*R*,9*S*,13*S*,14*S*)-3-hydroxy-13-methyl-7,8,9,11,12,14,15,16-octahydro-6*H*-cyclopenta[*a*]phenanthren-17-one. A somewhat less rigorous system names steroids as derivatives of a set of hypothetical steroid hydrocarbons. Steroid chemists, it should be noted, often use watered-down versions of the hydrocarbon system and in addition use colloquial names that are accepted by other scientists working in the same field. As an example, steroids lacking the 19-methyl group are more often designated 19-nor compounds than derivatives of the theoretical gonane hydrocarbon, The names that are used in this volume generally follow the earlier IUPAC convention based on the hypothetical hydrocarbons.

The nuclei that serve as templates for naming steroids are depicted in Scheme 1. The four rings are commonly denoted by the capital letters A, B, C and D reading from left to right. The ring letter designation, it should be noted, is used only in discussion sections of publications as they are have no role in formal nomenclature. The apparently eccentric numbering system used to denote steroid carbon atoms traces back to the days of the major structure determination work and reflects the then uncertainty about the overall structure (see Chapter 1). As noted previously, systematic names for steroids are based on a set of hypothetical hydrocarbons. Omitting the methyl group 19 at position 10, for example, affords gonanes (**1-1**) as noted above, more commonly known to practitioners as 19-nor steroids. This nucleus, as noted, serves as the base for several oral contraceptives. Estranes comprise compounds in which ring A is aromatic (**1-2**); these comprise an important part of oral contraceptives. Androstanes, the compounds that support male reproductive function, include a methyl group at position 10 (**1-3**). The other so-called sex hormones, the pregnanes (**1-4**), retain the intact 19 carbon atom nucleus and in addition sport a two-carbon side chain at position 17. The glucocorticoids, best know for their anti-inflammatory activity, are also named as derivatives of pregnane. Nucleus **1-5** (cholestane) depicts the most generalized structure that serves as the base for molecules with larger side chains at position 17, such as cholesterol and ergosterol and their derivatives.

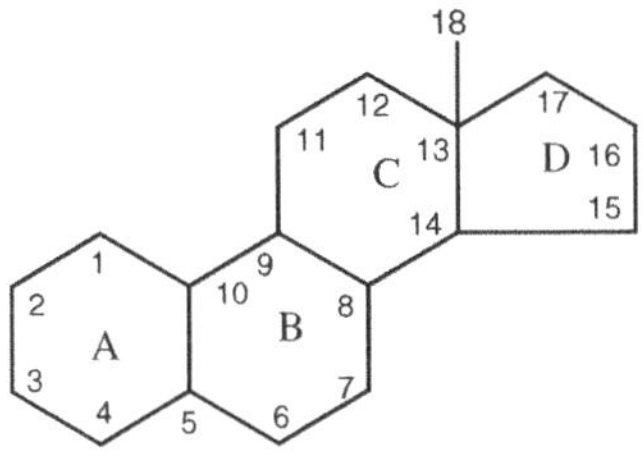

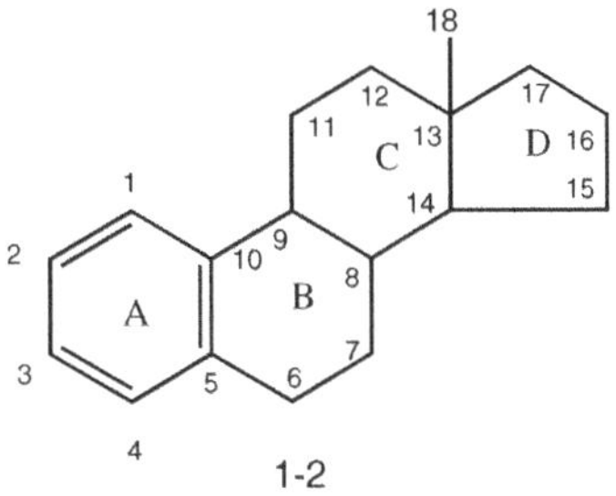

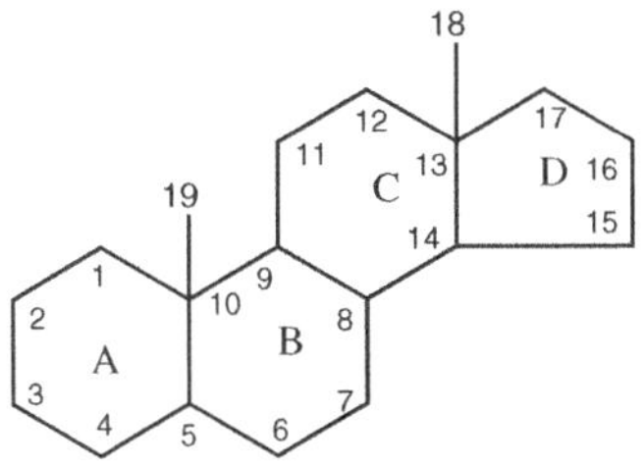

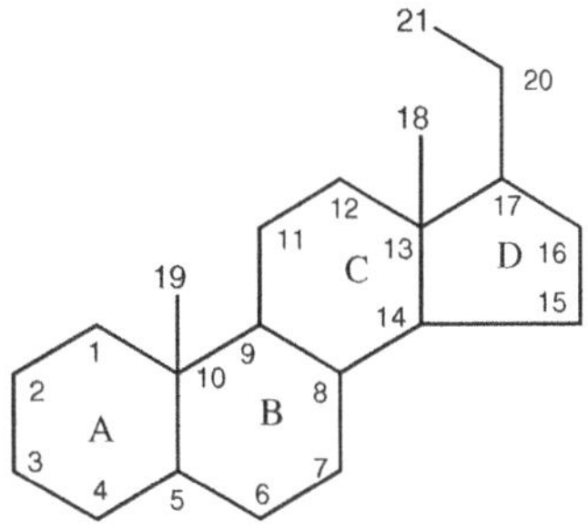

Scheme 1

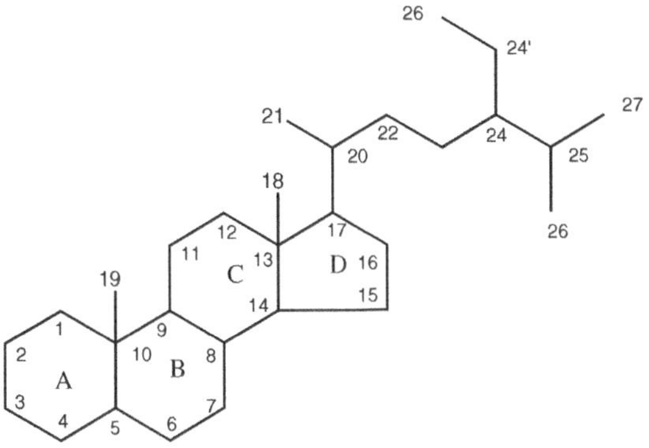

Standard depictions for steroids, such as those in Scheme 1, overlook the three-dimensional nature of the molecule. The structures in Scheme 2 represent the stereochemical arrangement implicit in the more customary formulas. The junction between rings B and C and also that between rings C and D involve transoid stereochemistry. It is of note in passing that the latter in fact comprises a disfavored *trans* hydrindane fusion. The ring junction between rings A and B is also transoid but can in selected cases assume a cisoid configuration (**2-2**). Note further that the stereochemical

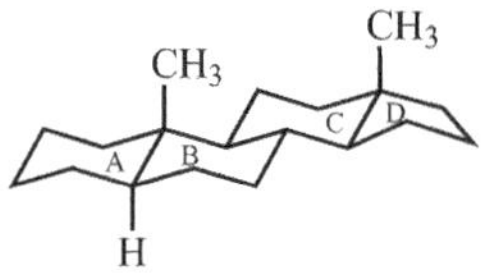

4 *Steroid Chemistry at a Glance*

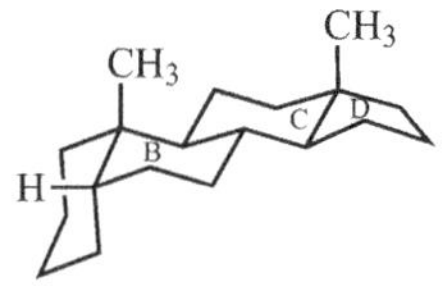

Scheme 2

depictions **2-1** and **2-2** in fact also present the absolute configuration of steroids from natural sources. This assignment is supported by detailed X-ray crystallographic studies. The founders of steroid chemistry can be said to have picked the correct depiction from a 50:50 choice. Substituents below the plane of the molecule are designated α, and those above are named β. The plane of the paper serves as the plane of the molecule for customary depictions.

Gonanes

Many if not most organic compounds are named as derivatives of some appropriate hypothetical hydrocarbon. This convention also holds true for steroids. Trivial names, other than those for 19-nor compounds, are uncommon in the gonane series, since these compounds have virtually no counterparts in Nature. The stereochemistry of the product to be named is assumed to be the same as that of the hydrocarbon. The arrangement is, however, made explicit for positions that can vary, for example, substituents at positions 3 and 5 in gonane **3-1** (Scheme 3). Proceeding systematically, compound **3-1** is a gonane with a hydroxyl at the 3-position oriented above the plane of the molecule, thus provisionally gonan-3β-ol. The hydrogen at carbon 5 can occur in either one of two orientations; the two resulting compounds are considered to be different systems for nomenclature. The hydrogen in this particular compound is oriented below the plane, assigning it to the 5Hα-gonane class. Compound **3-1** is thus named 5Hα-gonan-3β-ol. Compound **3-2** differs from the foregoing in the *cis* rather than *trans* fusion of rings A and B, with the result that hydrogen a position 5 is now β. This steroid is thus named 5Hβ-gonan-3β-ol.

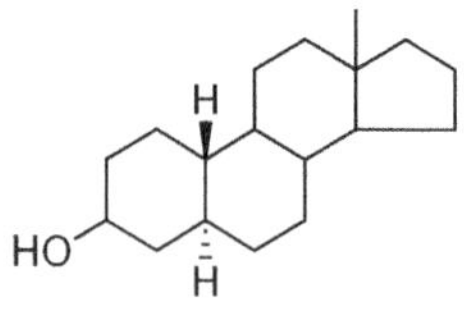

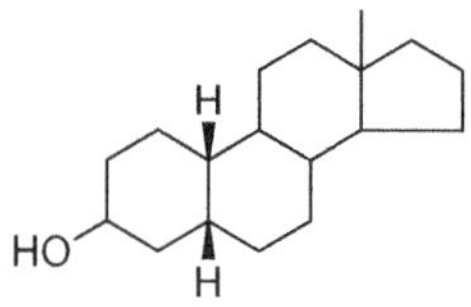

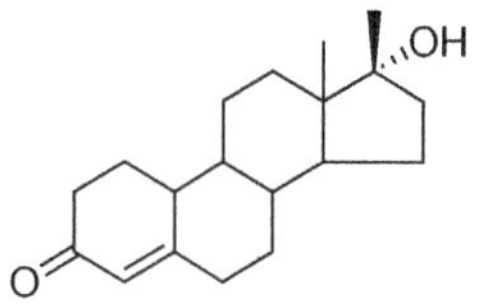

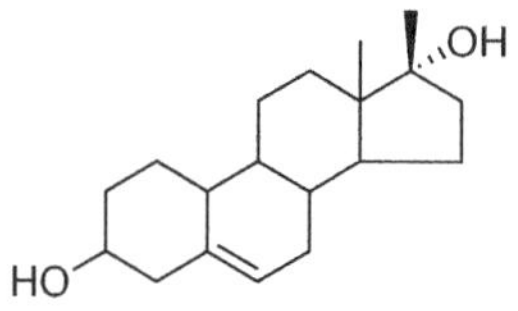

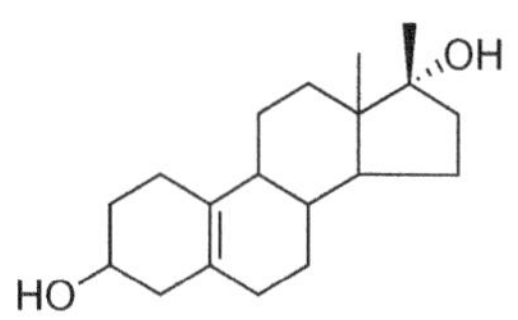

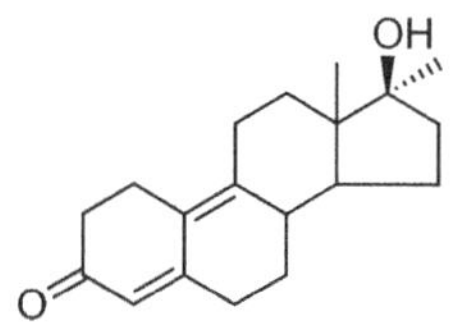

Scheme 3

Carbonyl groups are denoted by 'one' as in simpler non-steroidal compounds. The name for **3-3** thus starts as gonan-3-one. A double bond is indicated as an 'ene' and numbered for the lowest carbon that encompasses the unsaturation. This changes the name to gon-3-en-4-one. Taking into account the substitution on carbon 17, we get 17β-methyl-17α-hydroxygon-4-en-3-one. Proceeding in the same manner, the name for **3-4** becomes 3β,17α-dihydroxy-17β-methyl-gon-5-ene. Both termini of a double bond; with the higher in parentheses, are indicated in the name when they cannot be numbered sequentially. The isomer of the foregoing in which the double bond includes both bridgehead carbon atoms will be named 3β,17α-dihydroxy-17β-methylgon-5(10)-ene.

The methyl group at position 13 is replaced by ethyl in a series of gonanes prepared by total synthesis. One approach for naming for **3-6** posits deleting the methyl group at 17 by the notation '13-nor' and then putting in place an ethyl fragment: 17β-methyl-18-nor-13-ethylgon-4-en-3-one. The preferred name in a 1989 IUPAC publication simply elongates the methyl group by one carbon and designates the new group by the term 'homo'; **3-6** thus becomes 18-homo-17β-methylgon-4-en-3-one.

Estranes

Estrogenic activity in mammals is mediated by several estranes. A good number of these compounds are better known by their trivial as opposed to systematic names. The estrane nucleus is fairly similar to a gonane in that the double bonds in ring A are actually enumerated. Structure **4-1** (Scheme 4) thus becomes estra-1,3,5-trien-17-one. This steroid is far better known as estrone. Its reduced counterpart **4-2** is named estra-1,3,5-triene-3,17β-diol. The more prevalent name for this compound is estradiol or, somewhat less commonly, β-estradiol, The isomer with the hydroxyl group below the plane of the paper is named estra-1,3,5-trien-17α-ol or α-estradiol. The systematic name for the methyl ether of estradiol becomes 3-methoxyestra-1,3,5-trien-17β-ol. The product **4-4** from alkylation at position 17 can be named 17β-methylestra-1,3,5-trien-17α-ol. That compound is also known as equilin, a name derived from the fact that it was first isolated from horse urine. An additional double bond as in **4-5** is simply added to three already present, thus estra-1,3,5,7-tetraen-17-one. Additional substituents on the aromatic ring as in **4-6** are simply enumerated. The systematic name for this compound thus becomes 3-methoxy-17α-methylestra-1,3,5-triene-2,17β-diol.

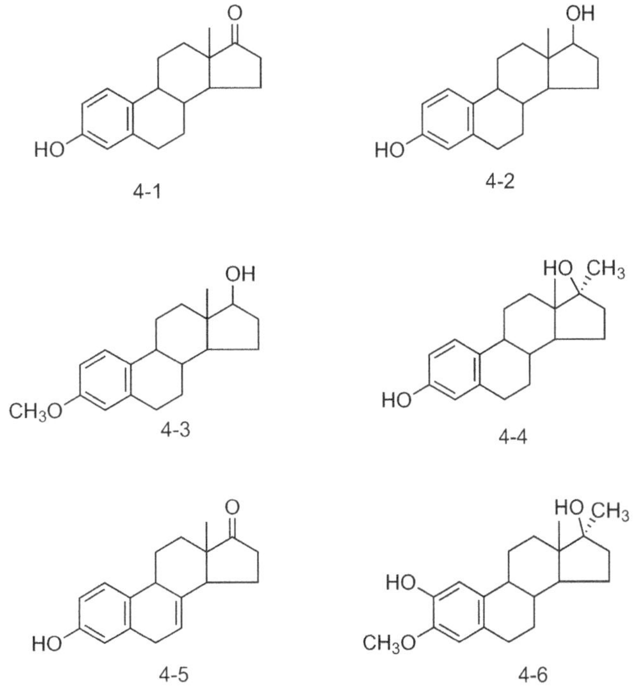

Scheme 4

Androstanes

Continuing the perusal of the steroid sex hormones leads to the series that has had extensive coverage in the popular press. The hypothetical hydrocarbon that forms the nucleus for this series features the full four-ring nucleus with methyl groups at both bridgehead carbon atoms. Applying rules similar to those used above, the formal name for compound **5-1** (Scheme 5) becomes 17β-hydroxyandrost-4-en-3-one. This steroid is again far better known by its trivial name testosterone. The double bond in testosterone is reduced *in vivo* to afford 5Hα,17β-hydroxyandrost-3-one (**5-2**), a compound that is significantly more potent than its precursor (**5-1**); this compound may well be the proximate hormone. The steroid from adding a methyl group at position 17 will be named 17α-methyl-17β-hydroxyandrost-4-en-3-one. Compound **5-4**, 3β-hydroxyandrost-5-en-17-one, is the widely used 'health supplement' dihydroepiandrosterone, abbreviated as DHEA. Substitution at position 6, as will be noted later, often increases potency. One such androgen, **5-5**, will be named 7α-methyl-17β-hydroxyandrost-4-en-3-one. Multiple

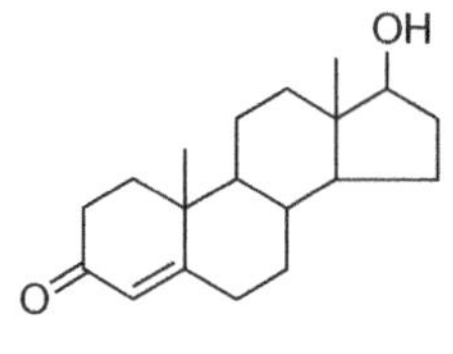

6 Steroid Chemistry at a Glance

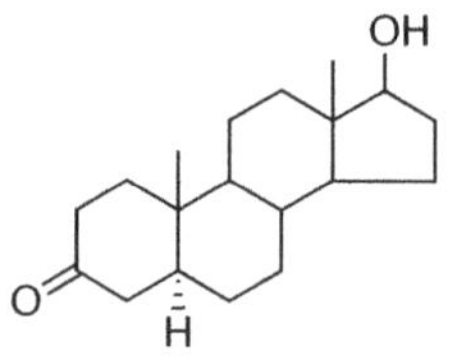

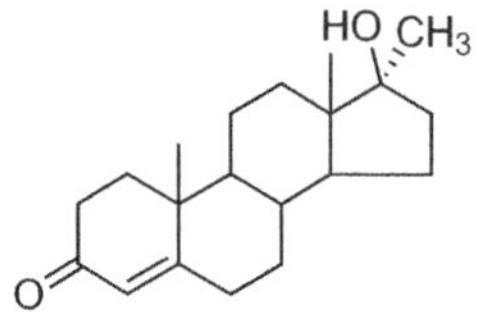

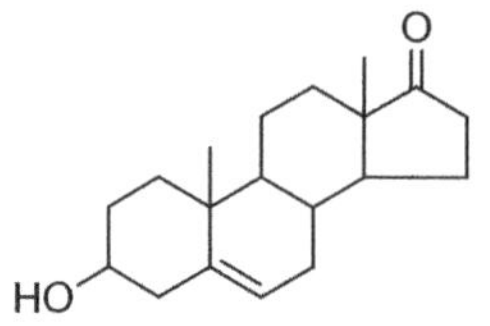

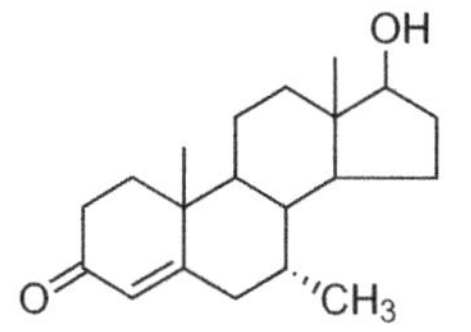

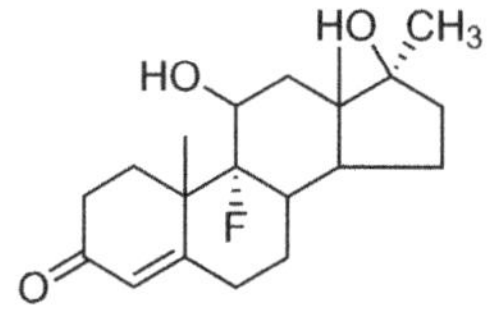

Scheme 5

modifications of testosterone are noted in increasingly complex names. The marketed steroid **5-6** thus becomes 9α-fluoro-11β,17β-dihydroxy-17α-methylandrost-4-en-3-one.

Pregnanes

The hypothetical pregnane hydrocarbon in essence comprises androstane with an additional two-carbon side chain at position 17. The default stereochemistry at that center places that side chain in the β configuration. The orientation of that group needs to be specified only for those cases where the side chain is below the plane of the molecule (α). A significant number of trivial names are used for these classes. Both progestins and glucocorticoids contain the two-carbon substituent at carbon 17. These two groups are treated separately due to the very large number of compounds in each category.

Progestins Taking the simplest compound first, note that the side chain in **6-1** (Scheme 6) occupies the β-position and therefore does not need to be specified. Incorporating changes on the hydrocarbon pregnane leads to pregn-4-ene-3,20-dione. This systematic name for **6-1** is, however, unlikely to displace the term progesterone. Much the same considerations apply to **6-2**. The trivial name for this compound, pregnenolone, is used far more widely than the systematic name for this steroid, 3β-hydroxypregn-5-en-20-one. Additional substitution at position 17 has been widely studied. The systematic name for one of these analogues, **6-3**, would be 17α-hydroxypregn-4-ene-3,20-dione rather than the widely used 17-hydroxyprogesterone. The more highly modified pregnane **6-4** becomes 6,17α-dimethylpregn-4,6-diene-3,20-dione. By the same token, **6-5** becomes 6β-methylpregna-1,4-diene-3,20-dione. The trivial name for **6-6**, 17-ethynyltestosterone, would probably be readily deciphered by one current in the field. The naming convention, however, holds that this steroid be designated as a pregnane, specifically 17α-ethynyl-17β-hydroxypregn-4-en-3-one.

Corticosteroids The pregnane carbon skeleton also serves as the nomenclature nucleus for a very large group of glucocorticoids. The discovery of the anti-inflammatory activity of cortisone (**7-1**) (Scheme 7) occasioned major programs in pharmaceutical company laboratories intended to develop better tolerated analogues and to increase potency. This resulted in the synthesis of compounds that carried modifications from cortisone on as many as half a dozen positions. All of those compounds, it should be noted, incorporated oxygen at carbon 11. As a general rule in building a name, modifications such as double bonds (ene) and carbonyl carbons (one) are treated as suffixes to the nucleus (pregn); other substituents such halogens, alkyl groups and hydroxyls precede pregn in alphabetical order. Compound **7-2**, comprises cortis*ol* with an additional unsaturation in ring A, a modification found in the great majority of modified corticosteroids. The pregnane-based systematic name for this steroid is 11β,17α,21-trihydroxypregn-1,4-diene-3,20-dione. The methyl group at C$_6$ on the modified corticoid **7-3** is attached to a trigonal carbon and thus does not need to be named. The systematic name of **7-3** is thus 17α,21-dihydroxy-6-methylpregn-1,4,6-triene-3,11,20-trione. Fluorinated

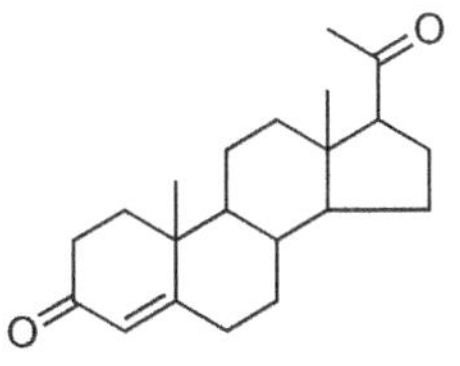

6-1

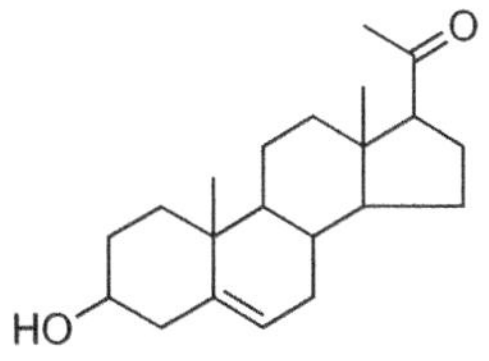

6-2

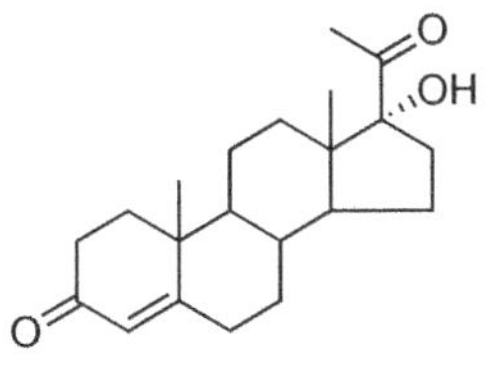

6-3

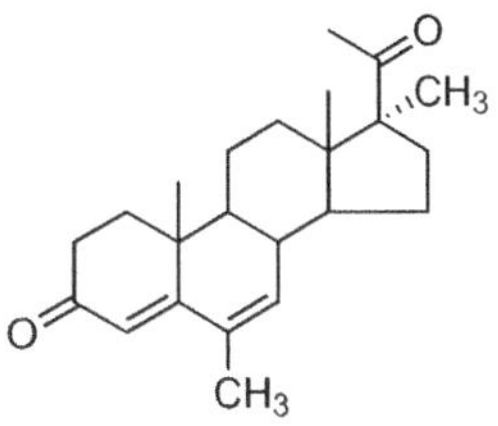

6-4

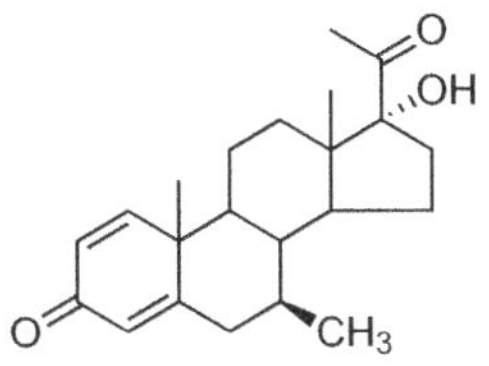

6-5

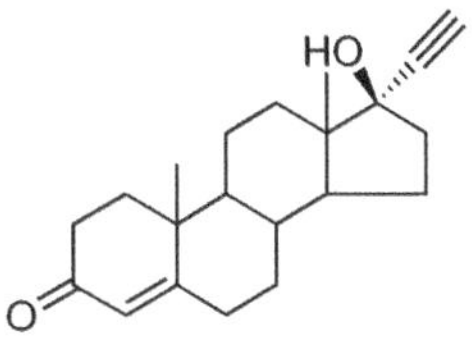

6-6

Scheme 6

cortisol **7-4** becomes 6α,9β-difluoro-11β,17α,21-trihydroxypregn-1,4-dien-3,20-one. Modification of ring D has led to some very potent compounds such as **7-5**; this analogue would be named 9α-fluoro-16α-methyl-11β,17α, 21-trihydroxypregn-1,4,6-trien-3,20-one.

Rigorous application of nomenclature to the final example, **7-6**, would treat the cyclic acetal as a fused dioxole. The alternative that names the compound as a derivative of the 16,17-diol is more common. The name thus becomes 21-chloro-9α-fluoro-11β,16α,17α-trihydroxypregn-1,4-dien-3,20-dione 16,17-acetonide.

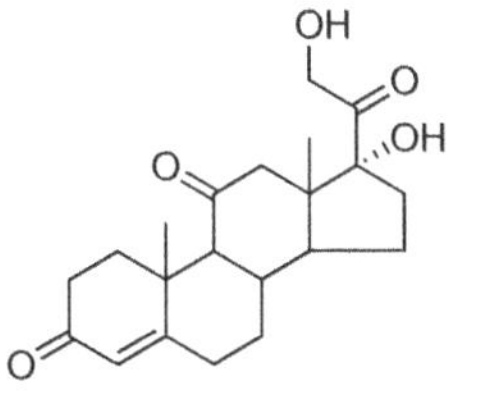

7-1

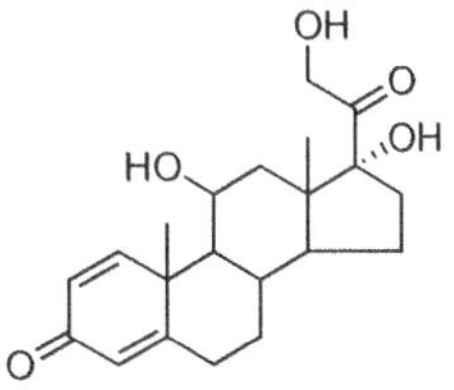

7-2

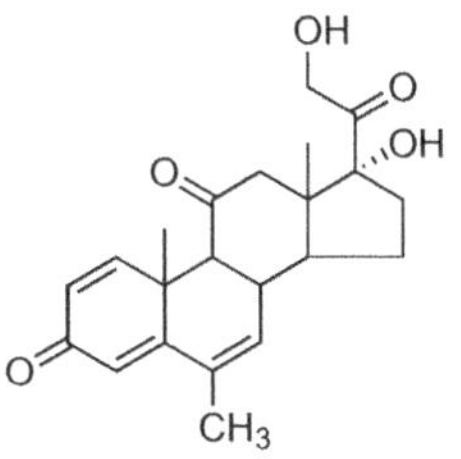

7-3

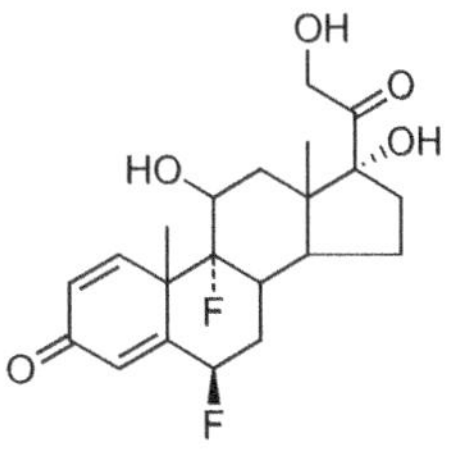

7-4

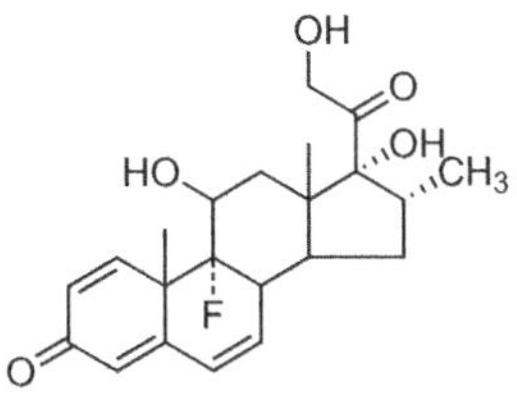

7-5

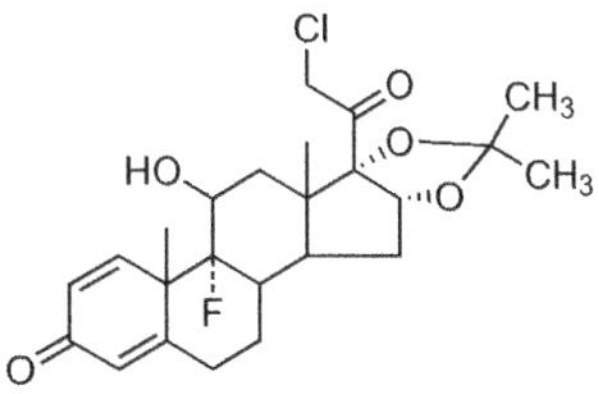

7-6

Scheme 7

The Bile Acids

As noted earlier, bile acids were among the first steroids to be obtained in pure crystalline form. These compounds played an important role in the effort devoted to divining the structure of steroids. Bile acids as a result acquired a sizeable number of trivial names, most of which gave little information as to their chemical structure. One approach to systematic names is based on the hypothetical cholanoic acid **8-1** (Scheme 8). Bile acids are then named as derivatives of this structure using the rules used for other classes of steroids. Note the *cis* A–B ring fusion in this series. The systematic name for **8-2**, lithocholic acid, is then simply 3α-hydroxy-5β-cholanic acid. Chenodeoxycholic acid, **8-3**, becomes 3α,7α-dihydroxy-5β-cholanic acid. The predominant acid in bile, **8-3**, is cholic acid itself, or, 3α,7α,12α-trihydroxy-5β-cholanic acid.

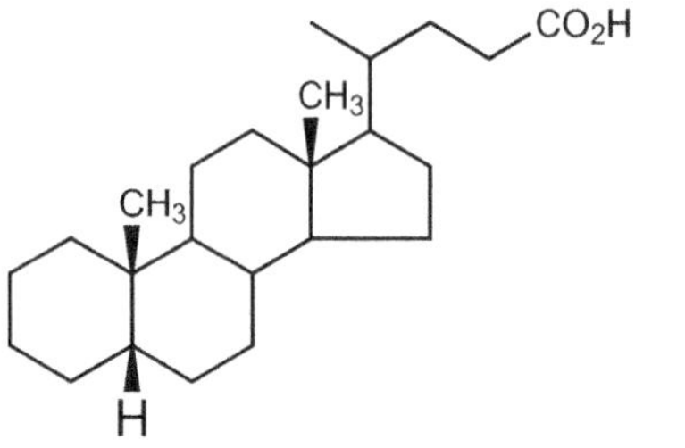

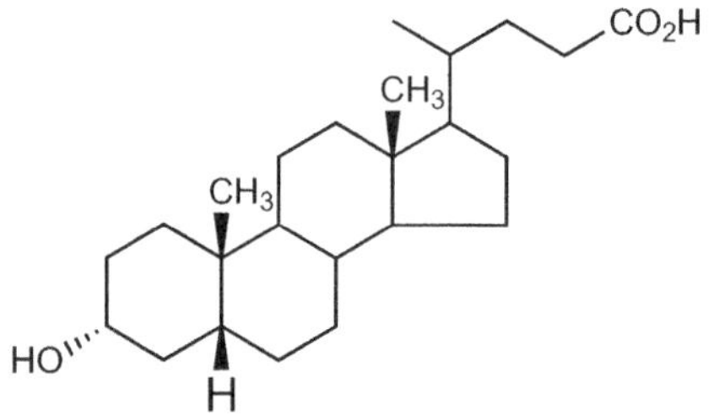

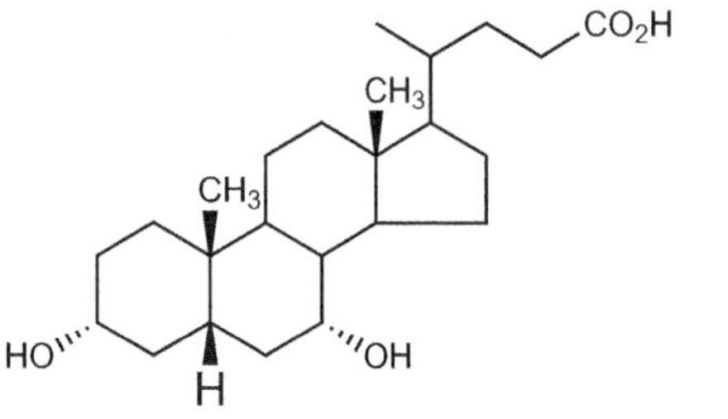

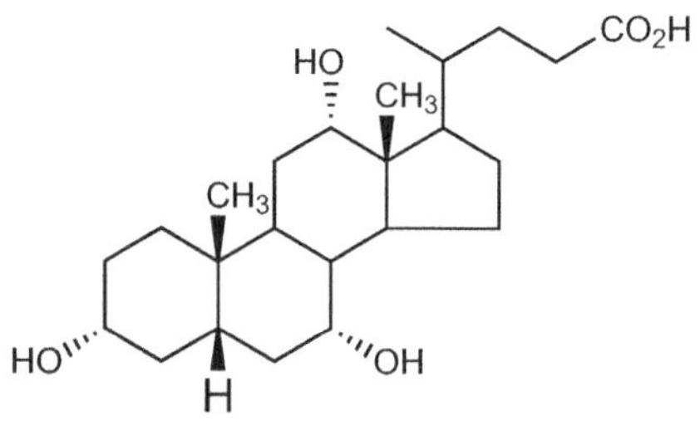

Scheme 8

Steroids with Ring Heteroatoms

Steroids in which a heteroatom replaces one of the ring carbons are named by inserting oxa or aza in addition to the position number in the name of the corresponding all-carbon compound. The closest all-carbon steroid to the anabolic agent **9-1** (Scheme 9) is 17β-androst-4-en-3-one. The name for **9-1** thus becomes 17β-hydroxy-17α-methyl-2-oxandrost-4-en-3-one. The cyclic lactam **9-2** can also be viewed as a modified androstane. Proceeding as in the previous example, **9-2** becomes 17β-(*tert*-butylcarbonyl)-4-azaandrost-3-one. Compound **9-3** can be related to

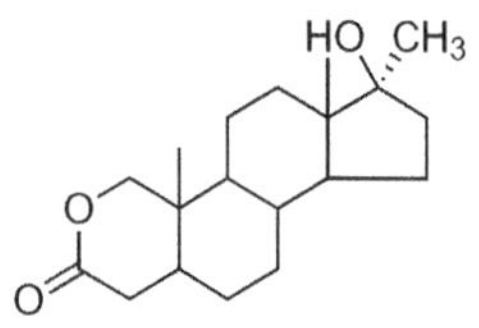

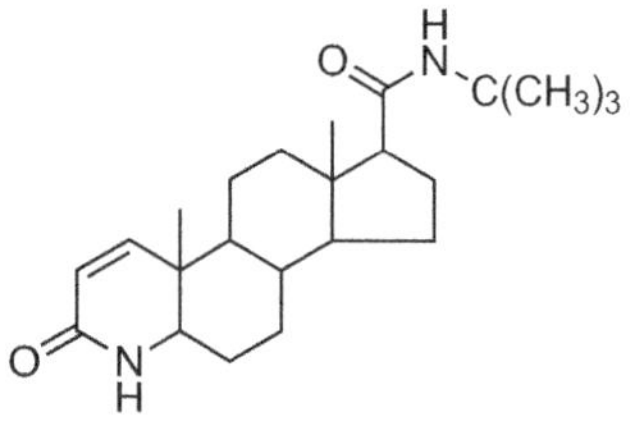

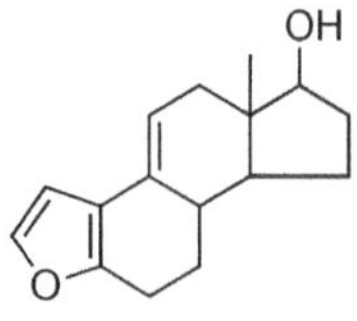

Scheme 9

estradiol in which one ring carbon atom is deleted while another has been replaced by oxygen. The compound is named after an estrane with omission of a ring carbon and is denoted by the term 'A-nor'. That structure is thus named 17β-hydroxy-A-3-oxanorestra-1,5(10),9(11)-triene.

Non-proprietary Names

The systematic names for many biologically active compounds are inarguably too complex for use in writing prescriptions. The notoriously bad physicians' handwriting would conspire to make matters worse. This is usually not a problem in the years when a compound is covered by patents since pharmaceutical companies tend to designate drugs by relatively short, easy-to-remember trade names. The other companies that market that drug once the patent has expired would then potentially have to designate the use of the systematic chemical name, presenting another hurdle piled on to that of legibility. This problem was addressed in the early 1950s by medical authorities in the USA and the World Health Organization (WHO) elsewhere. Each of the groups set forth a mechanism for generating relatively short names that gave some information about both the chemistry and clinical activity of drug substances. In practice, a non-proprietary name, more commonly called generic, is first coined by the sponsor. That name is then proposed formally to either the USAN (United States Adopted Name) in the USA and WHO elsewhere. These authorities examine the proposal as to whether it follows the published guidelines. These include proper use of class suffixes and arcana such as the one that proscribes the use of letter combinations that have different pronunciations around the world. ('Ph' is changed to 'F' and 'th' to 't'). The system for formally assigning names is administered in the USA by the USAN, an organization supported by the American Medical Association and the US Pharmacopoeia. International Non-proprietary Names (INNs) are overseen by the WHO. The system works fairly well for relatively new classes of drugs with similar structures or biological activities. Names for beta-blockers, for example, carry the suffix *-olol* and quinolone antibacterial agents end in *-acin*. The USAN/INN terms for androgens, for example, generally end in *-one*; the group of glucocorticoid steroids have several different suffixes with *-asone* and *-olone* predominating. The non-proprietary names that that will be found in this text do not always follow that system, since many of the drugs referred to precede the USAN and INN conventions.

Biological Activity of Steroids

The biological activity manifested by of each of the classes of steroids is described at the beginning of the relevant sections in this book. Those varied activities are mediated, however, at the single cell level by roughly the same mechanism.

The process whereby the great majority of the effector molecules that regulate vertebrate function begins with the binding of those compounds to the cell surface aspect of transmembrane receptors. This binding at the outer face of the cell membrane will, in one category of hormones, result in the release of secondary messengers, such as adenyl cyclase or leukotrienes, from the inner surface of the membrane. These newly released compounds then start the chain of reactions that lead to the measurable response. Alternatively, binding of a hormone or drug to surface receptors will open ion channels in the membrane; the influx or efflux of electrolytes will then lead to some observable event such as increased heart rate. These processes have in common the fact that the trigger molecule does not enter the cell; the overall result of binding is thus mediated by remote control.

Steroids, on the other hand, act entirely within the cell nucleus. These compounds cross cell membranes fairly rapidly, a process perhaps attributable to their lipophilic character. Once in the cytoplasm, the steroid may bind as administered or undergo some prior chemical changes. The steroid or its transformed product then binds to nuclear receptors present in intracellular fluid specific for that steroidal hormone. Some classes of steroids, such as estrogens, form complexes with nuclear estrogen receptors. The steroid–receptor complex then enters the cell nucleus; there the ligand–receptor complex interacts with a region on cellular DNA specific for that particular hormone. The binding event then triggers the chain that results in the synthesis of new protein. This overall process is manifested by the presence of new polypeptides. The standard laboratory assays for sex hormone activity have for many years comprised tests that measure growth in weight of sex-related tissues. Administration of estrogen to newborn female mice, for example, results in an increase in the weight of the oviduct as compared with untreated animals. An analogous experiment in male rats stimulates growth of the testes. In some cases, it should be noted, binding results in the synthesis of inhibitory peptides. Chronic administration of glucocorticoids results in wasting of muscle mass.

1
Steroids: a Brief History

The early history of steroids devolves almost exclusively about two compounds that had, at the time, been known for decades. These substances, cholesterol (**1-1**) and cholic acid (**1-2**) (Scheme 1.1) are available in large quantities from natural sources. This is arguably explains why these compounds were the first steroids to be obtained in pure crystalline form. Some gallstones, in fact, consist of as much as 90% of the neutral steroid cholesterol. This compound was isolated from those stones as pure crystals well before the birth of organic chemistry. The empirical formula for cholesterol, $C_{27}H_{46}O$, was established as early as 1888 (or alternatively $C_{54}H_{92}O_2$, as the concept of a molecule was at the time still somewhat nebulous and not accepted by all chemists). This compound and many of its derivatives, referred as sterols [from the Greeks *steros* (solid)] are also available from plants. Ox bile from slaughterhouses proved to be a relatively abundant source of bile salts. The acids from acidification of the salts consist largely of cholic acid (**1-2**) and chenodeoxycholic acid (**1-3**). Each of these compounds was obtained as pure crystals at about the same time as cholesterol. The bile acids, it was subsequently found, are formed in the liver by oxidation of cholesterol. They serve as surfactants for absorption of fats from the intestine and also for excretion of cholesterol and other hydrophobic compounds. The relative abundance of pure cholesterol and bile acids focused early research aimed at unraveling the chemical structure of steroids on those two compounds. This research actually preceded the discovery of the hormonal steroids by a good many decades. In view of their lack of biological activity, the investigations aimed at elucidating the structure of cholesterol and of cholic acid was probably undertaken largely as an exercise in structural organic chemistry. Results from these studies markedly facilitated subsequent efforts to assign structures to the so-called sex steroids.

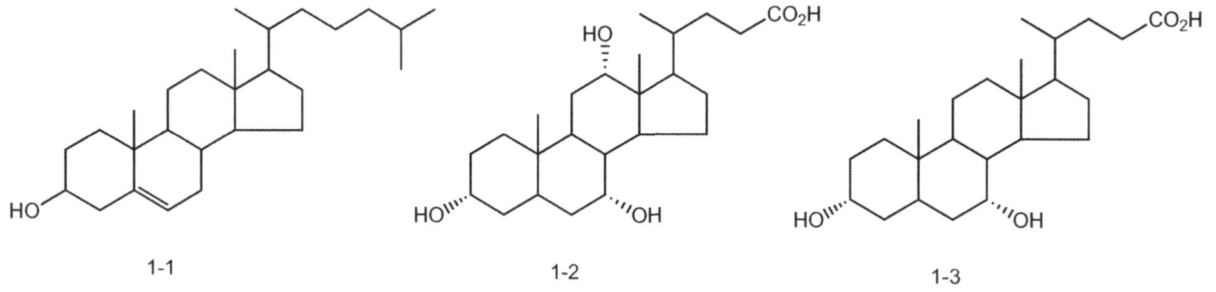

Scheme 1.1

1.1 Structure Determination

1.1.1 Cholesterol and Cholic Acid

The research aimed at determining the chemical structure of steroids long predated the availability of the instruments that form the backbone of today's work on the determination of the structures of natural products. Although the concept of infrared absorption had already been proposed in the mid-1920s, instruments for determining spectra would not be available until several decades in the future. The phenomenon of nuclear magnetic resonance (NMR) was unknown even to theoretical physicists. Had the concept been proposed, the use of that tool in structural studies awaited the invention of magnetron vacuum tubes as sources of microwave radiation (development of that electronic device was as a direct product of wartime (World War II) research on radar). In the first half of the 20th century, work on structure determination instead relied largely on degradation reactions that would reduce the target to ever smaller molecules until they matched compounds of known structure. The work also relied extensively on combustion analysis for determining elemental composition and Rast molecular weight determinations. Isolation of discrete products from degradation reaction mixtures required great technical skill in those days before the advent of any sort of chromatography. Elegant chemical reasoning played a very large role in interpreting the results of degradation experiments. As an example, the major product from heating the triene **2-1** (Scheme 1.2), likely obtained from dehydration of cholic acid (**1-2**), consists of the hydrocarbon chrysene (**2-2**), the structure of which had by then been independently established. This result provided early evidence for the presence in steroids of a staggered array of four fused rings.

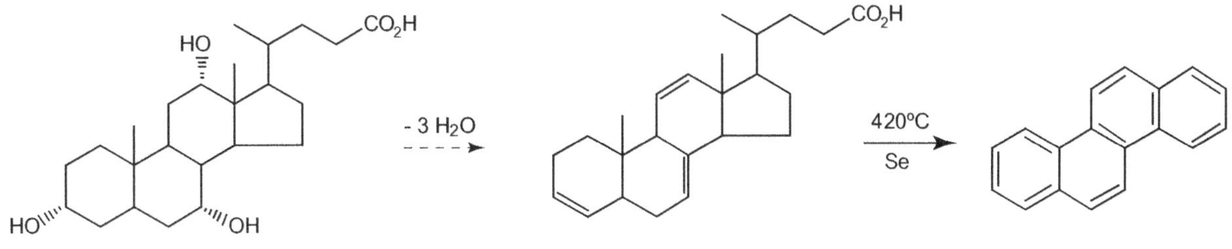

Scheme 1.2

More direct evidence for the gross structure of carbon skeleton of steroids came from the isolation of the hydrocarbon **3-1** (Scheme 1.3) from a mixture of hydrocarbons obtained from heating cholesterol itself with selenium. This product **3-1**, known as the Diels hydrocarbon, proved to be identical with a sample of the compound synthesized from starting materials of known structure by an unambiguous reaction scheme.

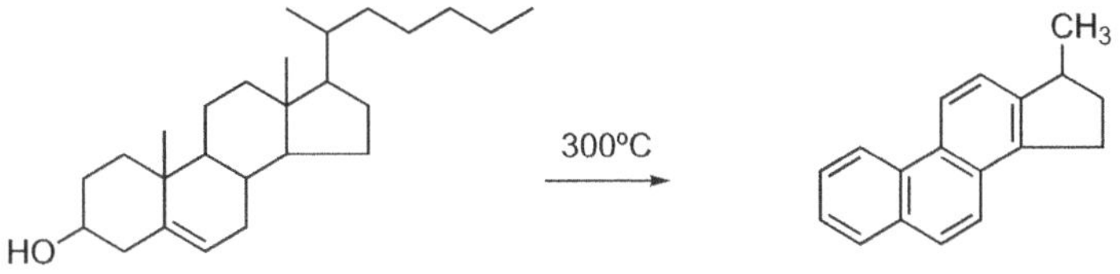

Scheme 1.3

The size of each of the rings present in steroids was established by serial oxidation reactions starting with what would later be dubbed ring A. The empirical, so-called Blanc, rule holds that oxidation of a cyclohexanone (**4-1**) (Scheme 1.4) proceeds to afford a dicarboxylic acid (**4-3**), likely through the enol form, **4-2**. Heating the diacid **4-3** with acetic anhydride proceeds to cyclopentanone **4-4** with loss of one carboxyl carbon. Repeated oxidation of that intermediate again results in a dicarboxylic acid, in this case adipic acid (**4-5**). Exposure to hot acetic anhydride leads to anhydride **4-6**. The strained nature of the cyclobutanone that would result from cyclization as in **4-3** is disfavored over the formation of the anhydride. Reaction thus proceeds to succinic anhydride (**4-6**). In the absence of instruments, anhydrides can be distinguished from ketones by the fact that the former will lead to a dicarboxylic acid on basic hydrolysis. The neutral ketone can be recovered unchanged under the same conditions.

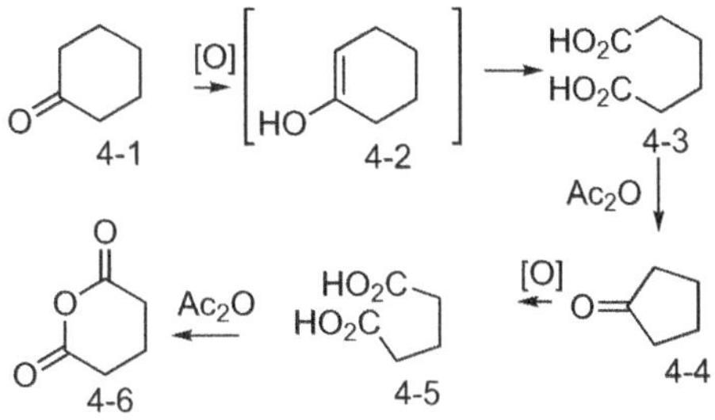

Scheme 1.4

In the case of the reduced derivative of cholesterol, **5-1** (Scheme 1.5), the initial oxidation goes to the highly substituted adipic acid **5-2**. The observation that this leads to a cyclopentanone (**5-3**) can be inferred to indicate that the ring at the start of the scheme was six-membered, Further oxidation of the cyclopentanone again leads to a dicarboxylic acid. On treatment with acetic anhydride, that intermediate leads to a cyclic anhydride. This leads to the inference that the precursor **5-3** was a cyclopentanone.

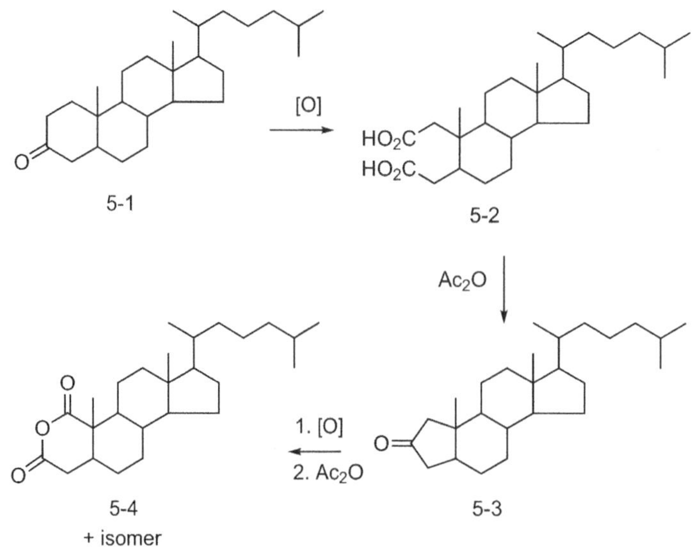

Scheme 1.5

Serendipity played a role in establishing the structure of the side chain in cholesterol. Some investigators had noted that a sweet, perfume-like odor accompanied the vigorous oxidation of cholesterol acetate (**6-1**) (Scheme 1.6). The odorous substance was finally isolated from a very large-scale (500 g) oxidation run and converted to its semicarbazone. This proved to be identical with the same derivative from 6,6-dimethylhexan-2-one (**6-2**).

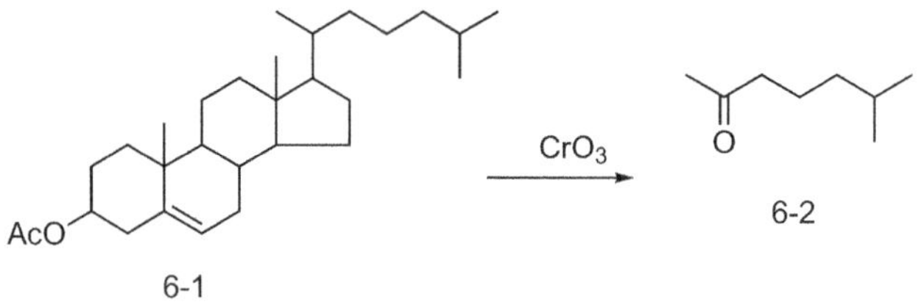

Scheme 1.6

Many chemists, principally those in Adolf Windaus's group at the University of Göttingen, worked on unraveling the intricacies of the structures in the cholesterol series. Another group, led by Heinrich Wieland at the University of Munich, studied the structures of the bile acids. Suspecting that these two natural products shared a common carbon nucleus, they each sought to relate the two by preparing a common derivative. In brief, they established that the cholanic acid **7-2** from exhaustive reduction of cholic acid (**7-3**) was identical to the product from oxidation of coprostane (**7-1**) (Scheme 1.7). The latter was obtained by exhaustive reduction of cholesterol. The common derivative, it should be noted, incorporates the less common cis A–B ring fusion. The reactions that lead to the common intermediate are unlikely to alter the configuration of chiral centers present in the natural products. The identity of the derivatives obtained from each starting material thus established that the product related to cholesterol and those derived from bile acids shared the same nucleus and overall stereochemistry.

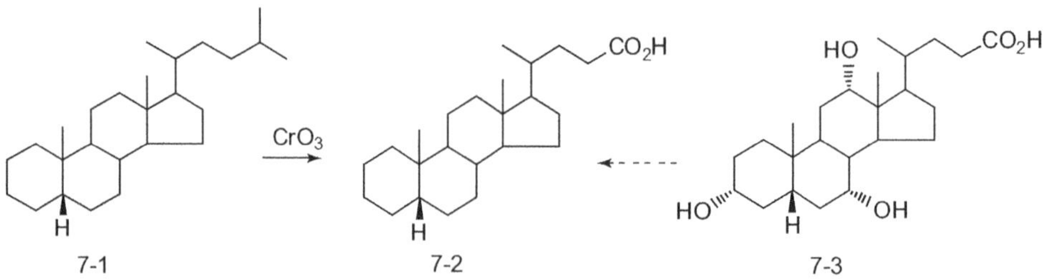

Scheme 1.7

The two groups and also other investigators who worked on the problem felt that enough data had been accumulated to propose a structure in 1928. Most of the carbon atoms had been accounted for and the results, they deemed, supported

8-1 (Scheme 1.8) as the structure of cholesterol. Some ambiguity existed as to the attachment of one of the two methyl groups in cholesterol. Some, it is said, referred to this fragment as the 'floating methyl'. Depiction of the proposed structure in three dimensions (**8-2**), instead of the common two-dimensional notation (**8-1**), makes it clear that the proposed structure would have consisted of a relatively thick, congested molecule.

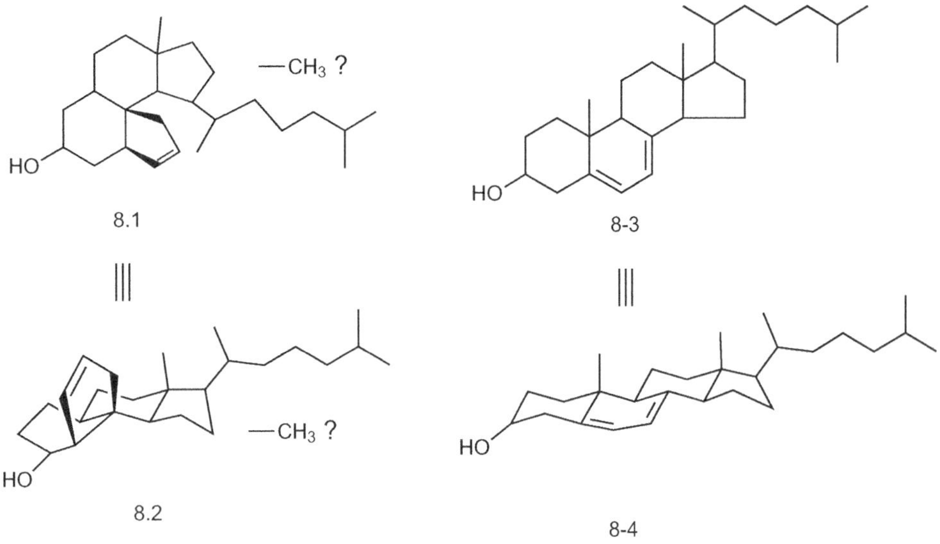

Scheme 1.8

The use of X-ray crystallography for solving the structures of organic compounds was still in its infancy in the late 1920s. The use of that tool was hindered by the need to perform an enormous amount of data reduction; the mechanical calculating machines employed for that were then just coming into wider use. Atom-by-atom mapping of a complex structure such as a steroid was, at that time, still beyond the then state-of-the-art. Resolution of an X-ray crystallographic study of ergosterol (**8-3**) was however sufficient to indicate that this steroid consisted of a long, flat molecule (**8-4**) rather than a thick, congested entity such as **8-2**. Re-examination of all the data from degradation studies revealed that an exception to the Blanc rule caused assignment of the wrong structure (**8-2**) to ergosterol. This second look also led to the correct formulation of the steroid nucleus as depicted by **8-3**.

One set of degradation studies on cholanic acid (**9-1**) (Scheme 1.9) led to scission of what are now known as ring A and ring C. One pair of the four new carboxylic acids led to a cyclopentanone and the other to an anhydride. On the basis of this, it was then inferred that ring A was six-membered whereas ring C comprised a cyclopentane (see also **8-1**). It was later recognized that carboxylic acids attached directly to rings as in **9-2** cannot form a cyclopentanone. This exception was later attributed to steric strain in the hypothetical product.

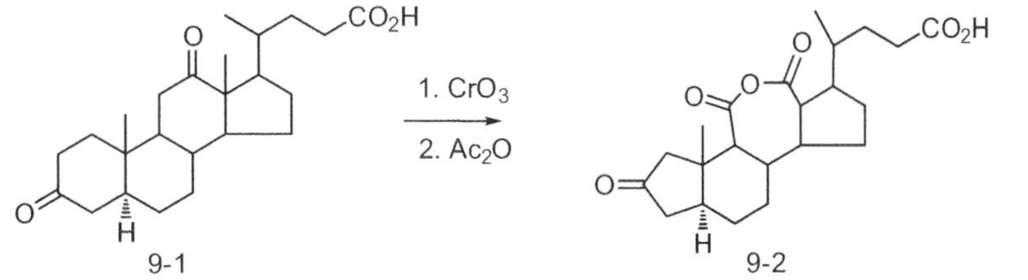

Scheme 1.9

1.1.2 The Sex Steroids

By the early 1930s, it was clear that the reproductive function in mammals was directed by a group of potent discrete chemical substances. These compounds, dubbed sex hormones, consist of three distinct classes, the estrogens, the progestins and the androgens; these substances differ from each other in both biological activity and structure. The very small amounts of these compounds found in tissues posed a major challenge to investigations aimed at defining their chemical structure.

1.1.2.1 Estrogens

The first of the three classes of hormones that regulate reproductive function in both females and males of the species, the estrogens, progestins and androgens, were isolated in 1929. This marked contrast to the dates for the first isolation of bile acids and of cholesterol is due in no small part to the minute amounts of those hormones that were available for structural studies. Isolation of those substances from mammalian sources, such as mare's urine, was guided by bioassays, increasing potency of a sample signaling higher purity. This work culminated in the isolation in 1929 of a weakly acidic compound, estrone (**10-1**) (Scheme 1.10). The acidity indicated the presence in the molecule of a phenol and hence an aromatic ring. Various chemical tests pointed to the steroid nature of estrone and also several closely related compounds. The principal accompanying compound, estradiol (**10-2**), and estrone comprise the primary estrogens and are freely intraconvertible both *in vivo* and *in vitro*. The former occurs as two isomers that differ at position 17; one isomer features the alcohol at position 17 above and the other below the plane of the molecule. Estradiol-β that carries the hydroxyl above the plane is the more potent than its 17α-hydroxy epimer. The closely related compound estriol (**10-3**) often accompanies estradiol *in vivo*. The compound can also be prepared from estradiol by a straightforward sequence of reactions not likely to change the carbon skeleton.

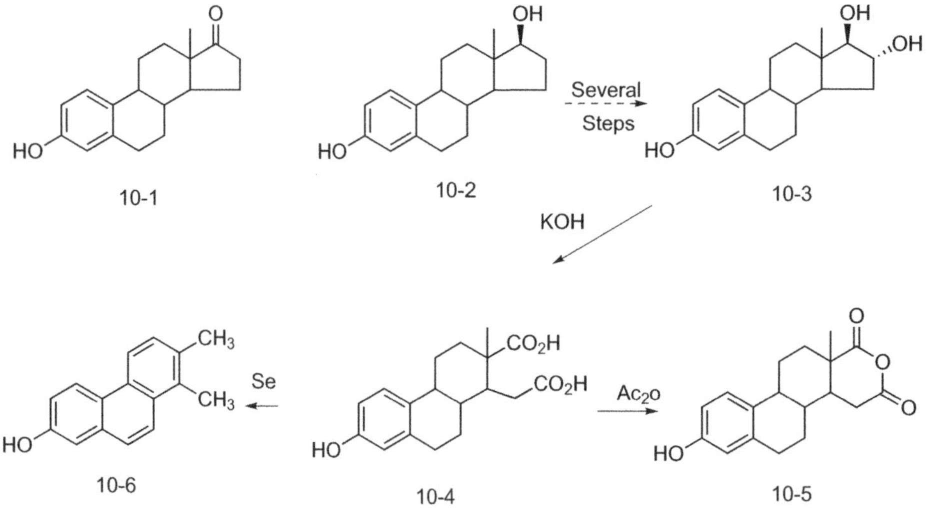

Scheme 1.10

Fusing estriol (**10-3**) with potassium hydroxide cleaves the bond between the two hydroxyl-bearing carbon atoms in the five-membered ring. Those atoms are oxidized to carboxylic acids under reaction conditions to afford **10-4**, dubbed marrianolic acid (this derivative, named after its discoverer, interestingly retains significant estrogenic activity). On treatment with acetic anhydride, this gives an anhydride, **10-5**; the Blanc rule indicates that the precursor ring is five-membered. Heating the diacid **10-4** with selenium causes the carboxyl groups to leave as carbon dioxide; under reaction conditions, the six-membered rings lose hydrogen, leaving behind the phenanthrol **10-6**. This molecule proved to be identical with a sample of **10-6** synthesized by an unambiguous route. By 1933, the detailed structure of estradiol was firmly established. Several total syntheses have been published since then (see Chapter 3).

1.1.2.2 Progestins

Biological studies carried out at roughly the same time identified another hormone, this one a neutral substance whose concentration in body fluids fluctuated in synchrony with the menstrual cycle; the hormone was further found in blood at high levels during pregnancy. This compound, progesterone, was not obtained in crystalline form until 1934 as it was often accompanied in extracts by other closely related compounds such as pregnenolone. Instrumental tools in this case played a role in deducing the chemical structure of progesterone. Although X-ray crystallography did not as yet enable atom-by-atom mapping, it did provide evidence that progesterone possessed a four-ring sterol-like structure. The ultraviolet spectrum showed an absorption spectrum typical for a conjugated unsaturated ketone. Much of the structural work was carried out using about 2 g of pregnanediol (**12-4**) obtained by extracting in excess of 1000 L of pregnancy urine.

A fairly straightforward scheme, called the Barbier–Wieland degradation, was at that time used for determining the number of carbon atoms in an aliphatic acid fragment. This involves first converting the carboxylic acid **11-1** to the corresponding ester **11-2** (Scheme 1.11). Reaction of the ester with phenylmagnesium bromide leads to carbinol **11-3**. Treatment with acid leads to dehydration and formation of the olefin **11-4**. Oxidation by one of several methods cleaves the double bond with formation of a new carboxylic acid (**11-5**), shorter by one carbon atom than the starting acid **11-1**. The sequence would be repeated until degradation met a branch in the chain and afforded a ketone instead of an acid.

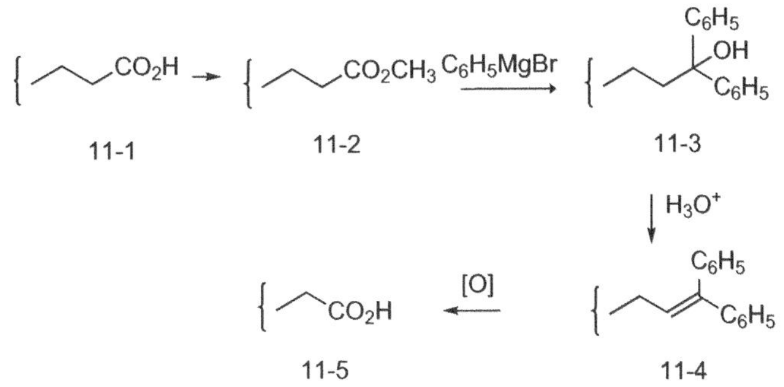

Scheme 1.11

The sequence that established the structure of the pregnan nucleus starts with the chain length probing sequence depicted in Scheme 1.12. The carboxylic acid derivative **12-1**, which can, in concept, be prepared from cholanic acid by initial exhaustive reduction to remove the hydroxyl groups followed by two rounds of sequence depicted in Scheme 1.11. The carbonyl group at position 20 (**12-3**) was reduced by means of amalgamated zinc to give the pregnane nucleus **12-6**. In a convergent sequence, pregnanediol (**12-4**) was oxidized to pregnane-2,17-dione (**12-5**) with chromium trioxide. The carbonyl groups at C_3 and C_{20} were then reduced with amalgamated zinc to give a sample of **12-6** identical in all respects with that obtained from cholanic acid (**12-1**). The assignment of one of the oxidized carbon atoms at position 3 was based on analogy with cholesterol and the other at position 20 relied on intra-conversion with the C_{17} androgens.

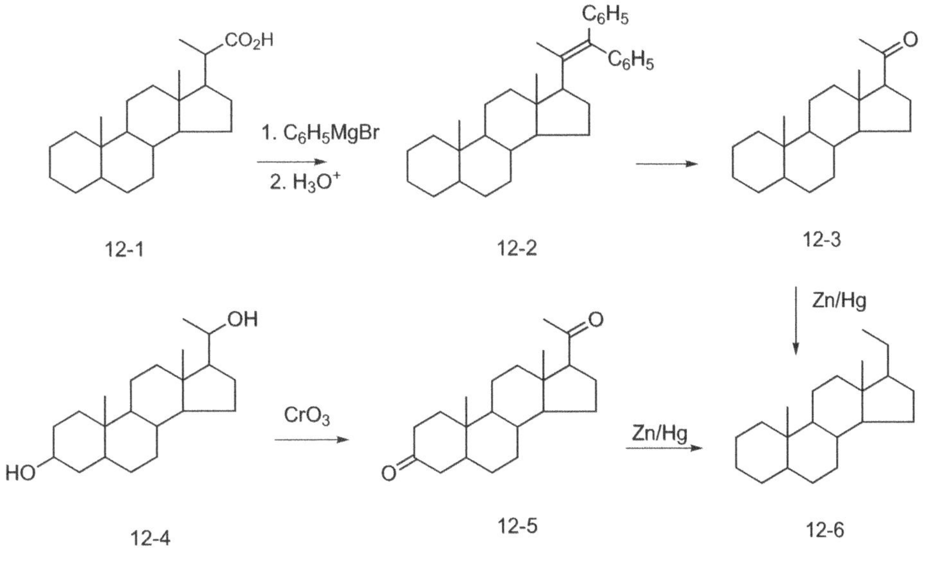

Scheme 1.12

1.1.2.3 C_{19} Androgens

Androgens, the male sex hormones, proved far more elusive that either the estrogens and progestins since they occur at much lower concentrations in biological fluids. The bioassay used to track the isolation in this case comprised the 'capon unit'. This was the amount of extract that produced a 20% increase in the surface of a rooster's comb. The 15 mg of pure crystalline testosterone isolated in 1931 came from about 15 000 l of urine. The structural investigations of this series relied on the then newly discovered side chain oxidations of cholestanol (**13-1**) (Scheme 1.13). This method in essence comprised fairly drastic oxidation of reduced cholesterols of known stereochemistry at the A–B junction to afford in fairly low yield products in which the side chain at C_{17} had been consumed to leave behind a carbonyl group. One of these products proved to be identical with androsterone (**13-2**). That compound had in turn been obtained from a sequence of reactions starting from dehydroepiandrosterone (**13-3**) that had been isolated from male urine.

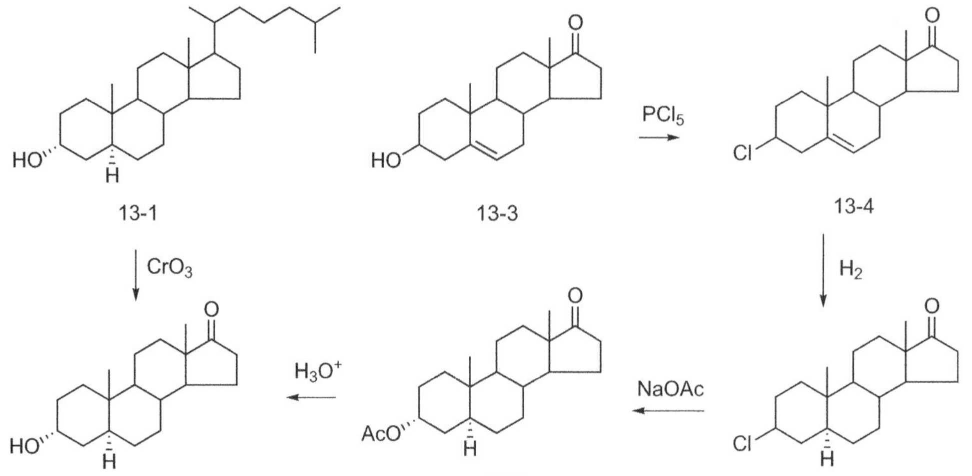

16 *Steroid Chemistry at a Glance*

Scheme 1.13

Treatment of dehydroepiandrosterone (**13-3**) with phosphorus pentachloride replaces the hydroxyl at position 3 with retention of configuration (**13-4**). It had been established prior to this work that catalytic reduction of unsaturation in steroids proceeds almost invariably from the bottom side to afford reaction products as their 5α epimers, as for example **13-5** [this is also the case for cholestanol (**13-1**), obtained from hydrogenation of cholesterol]. Sodium acetate then displaces chlorine in **13-5** to afford the acetoxy derivative **13-6** with inverted configuration at C_3. Mild hydrolysis of the acetoxy group affords the corresponding alcohol **13-2** that is identical with that of a sample produced by oxidation of cholestanol (**13-1**).

The stereochemical argument can be closed with the observation that oxidation of dehydroepiandrosterone by the Oppenauer reaction (aluminum isopropoxide in the presence of a ketone) yields the oxidation product androst-4-ene-3,17-dione (**14-1**) (Scheme 1.14). The same diketone is formed from oxidation of testosterone (**14-2**). Going in the reverse direction, androst-4-ene-3,17-dione can be converted to testosterone by treatment with fermenting yeast.

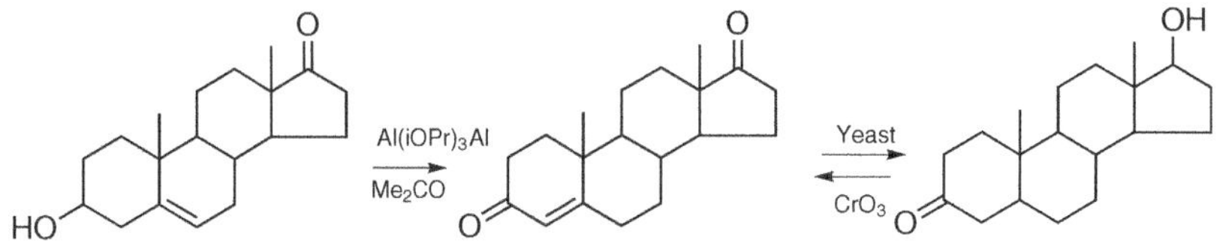

Scheme 1.14

1.1.3 Corticosteroids

1.1.3.1 Glucocorticoids

The realization in the 1930s that substances secreted by the so-called endocrine glands play a major part in various life processes led to a major effort to determine the chemical structure of those secretions, as was the case for the set of sex hormones described in the preceding sections. Investigation of the adrenal glands, located atop the kidneys, revealed that animals whose adrenals had been removed died within a few days. Administration of adrenal extracts increased their lifespan. Identification of the active ingredient was complicated by the 30 or so compounds, now known to be steroids, secreted by the adrenal outer layer, known as the cortex. The principal products, hydrocortisone (formerly cortisol) and aldosterone, account for most of the activity of the extracts. Hydrocortisone regulates carbohydrate, fat and protein metabolism whereas aldosterone acts on electrolyte balance via the kidneys. These hormones, like the sex hormones, occur at low concentrations: about 450 kg of beef adrenals yielded only about 300 mg of hydrocortisone. One of the compounds accompanying hydrocortisone proved to be identical with 20-hydroxyprogesterone (**15-6**) (Scheme 1.15) that had been prepared from the known carboxylic acid **15-1** in studies on the structure of progesterone.

The sequence for preparing the hydroxyketone started by conversion of the acid **15-1** to its chloride with thionyl chloride. Reaction of that acid halide with diazomethane gives the diazoketone **15-2**. The hydroxyl group at C_3 is then oxidized to the corresponding ketone by means of an Oppenauer reaction. Treatment of the product **15-3** with gaseous hydrogen chloride replaces nitrogen in that intermediate by chlorine. Displacement of chlorine by acetate then leads to the 21-acetate **15-5**. Saponification of the ester completes the sequence.

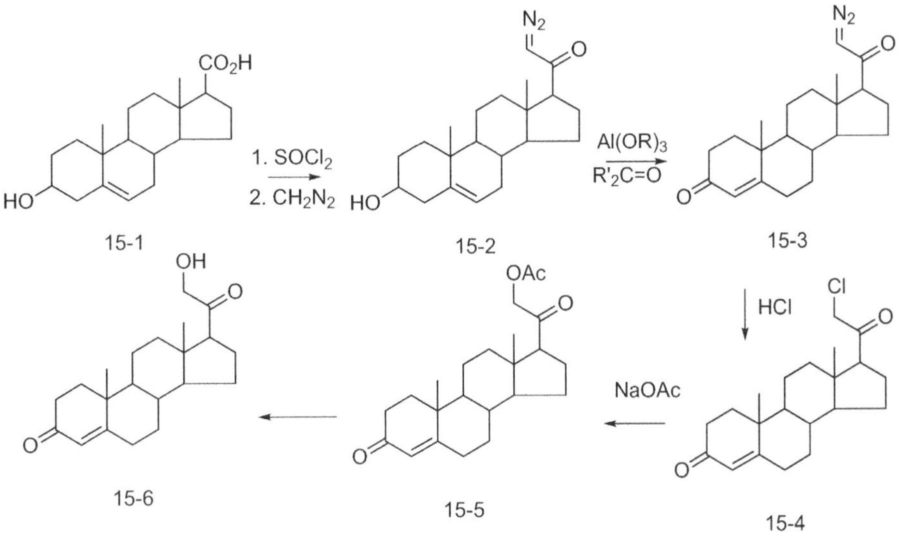

Scheme 1.15

Elemental analysis of hydrocortisone indicated the presence in the molecule of two additional oxygen functions compared with progesterone. Treatment of hydrocortisone **16-1** with periodic acid cleaves the side chain at position 20 to afford hydroxy acid **16-2**, a reaction that indicates the presence of a 1,3-dihydroketone (Scheme 1.16); this is also evidence that one of those additional oxygen atoms is at C_{17}. Further oxidation of **16-2** with chromium trioxide then causes the hydroxy acid to lose carbon dioxide to leave behind a ketone (**16-3**). The hydroxyl group at position 11, which is virtually inert to other reactions, is oxidized to a ketone under these reaction conditions; that product, known as adrenosterone, can also be formed by direct oxidation of **16-1** with chromium trioxide. Reduction of the double bond in **16-3** followed by treatment of the product with zinc amalgam in acid gives the hydrocarbon androstane (**16-4**). This proved to be identical with a sample prepared from dehydroepiandrosterone. It might be noted in passing that androstane emits a very strong odor, similar to that of poorly maintained urinals.

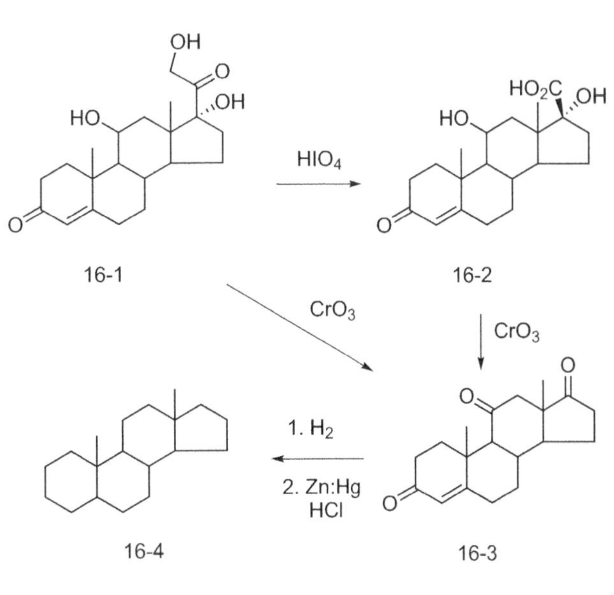

Scheme 1.16

Assignment of the remaining hydroxyl group to position 11 rests in large part on the lack of reactivity of the hydroxyl group in hydrocortisone (**16-1**), or for that matter the 11-ketone (**17-1**) in cortisone (Scheme 1.17). Molecular models show that the 18- and 19-methyl groups effectively shield those positions from attack from the β side. The ketone at position 11 will form normal derivatives only under the most forcing conditions. Reduction of the suitably protected form of the ketone readily gives the corresponding hydroxyl (**16-1**). This is assigned as β on the basis of the approach of hydride or hydrogen from the more accessible α face of the steroid. Additional support for placing oxygen at position 11 comes from the finding that dehydration results in the formation of a double bond at position 9(11) (**17-2**), whereas the 12-hydroxyl in 12-hydroprogesterone gives an 11,12-olefin (**17-3**).

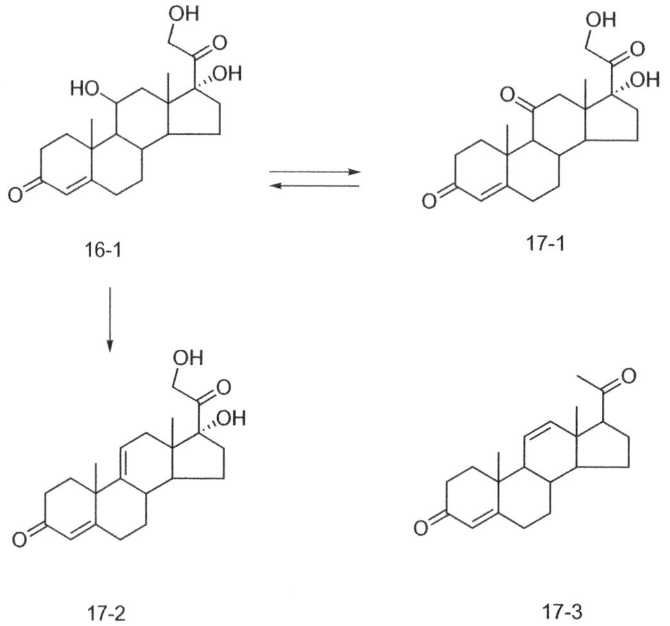

Scheme 1.17

1.1.3.2 Aldosterone

The work that led to the identification of cortisone in extracts of the adrenal cortex led, as noted above, to the isolation of a host of closely related steroids, There remained, however, a fraction that defied crystallization. This material, the amorphous fraction, exhibited fairly respectable activity in regulation blood volume and serum electrolytes. This steroid, aldosterone, can exist in either the keto or lactal form (Scheme 1.18). Degradation and synthesis studies are beyond the scope of this book.

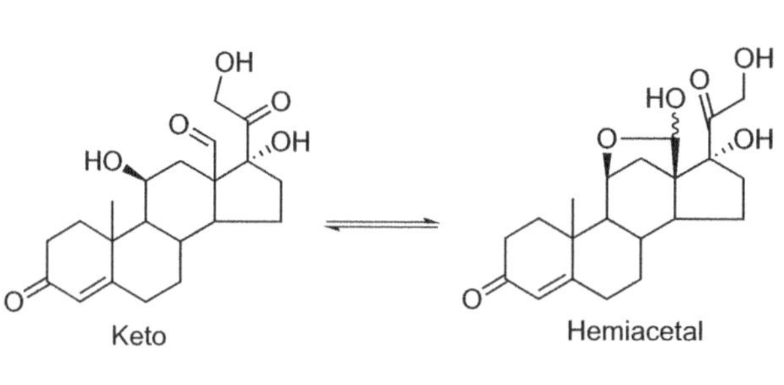

Keto   Hemiacetal

18

Scheme 1.18

2
Sources of Steroids

2.1 Biosynthesis

The activated derivative of isoprene pyrophosphate (IPP), as noted in the Introduction, comprises the five-carbon starting synthon for a host of hydrocarbon-like compounds that occur in both animals and plants. The term terpene is customarily used to denote the class of compounds made up of two isoprenes. Triterpene thus comprises compounds built from six isoprenes or 30 carbon atoms. That leads to an enormous number of possible structures since there are few restraints on the manner in which IPP couples. Literally hundreds of 30-carbon triterpene natural products have thus been isolated and characterized. Only one member of that group, lanosterol (**4-1**), is relevant to this discussion of steroids.

In rough outline form, one route to the isoprene derivative IPP starts by addition of an activated acetyl unit to activated acetoacetate **1-1** to form the glutarate **1-2** (Scheme 2.1). The enzyme HMG-CoA-reductase (3-hydroxy-3-methyl-glutaryl-CoA) then reduces the thiocarbonyl group in **1-2** to the corresponding alcohol to afford mevalonic acid (**1-3**) (shown as its phosphate-free acid). It is of interest to note in passing that the widely prescribed cholesterol-lowering statins act by inhibiting that specific enzyme. Although the structures of individual statins vary significantly, all incorporate a mevalonate moiety in their structure.

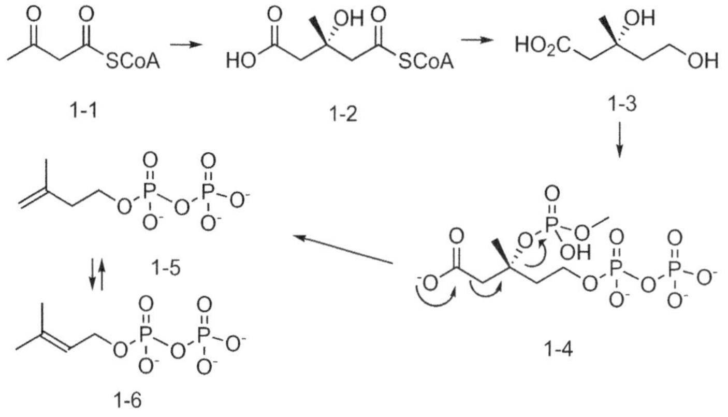

Scheme 2.1

The hydroxyl groups in mevalonate are then esterified in a stepwise fashion to their phosphates by phosphorylating enzymes, called kinases. The overall result is the intermediate **1-4** in which the terminal hydroxyl is a diphosphate and the tertiary hydroxyl is present as a monophosphate, the latter comprising a very good leaving group. The next step consists of decarboxylation with concomitant departure of the tertiary phosphate group. This transformation finally affords the latent isoprene unit **1-5**. The product from this sequence, IPP (**1-5**), is reversibly converted to its isomer with the internal double bond (**1-6**) by yet another enzyme.

As a general rule, natural products derived from isoprenoid units arise from head-to-tail reactions of that synthon. The majority of the structures of such products built up by condensation of IPP will as a result display a branched methyl on every fifth atom in the chain. Thus, reaction of IPP (**1-5**) with the isomer with the internal double bond (**1-6**) proceeds by head-to-tail coupling with expulsion of a pyrophosphate ion. The free alcohol from the product **2-1** (Scheme 2.2) is the fragrant terpene geraniol. Reaction of **2-1** with a second isoprene unit in this case again takes place by head-to-tail reaction to afford **2-2**. The free alcohol from this 15-carbon triene is known as farnesol and is generally classed as a sesquiterpene (Latin *sesqui-*, one and a half). The molecule is displayed in the unlikely conformation **2-2b**, in anticipation of the next reaction.

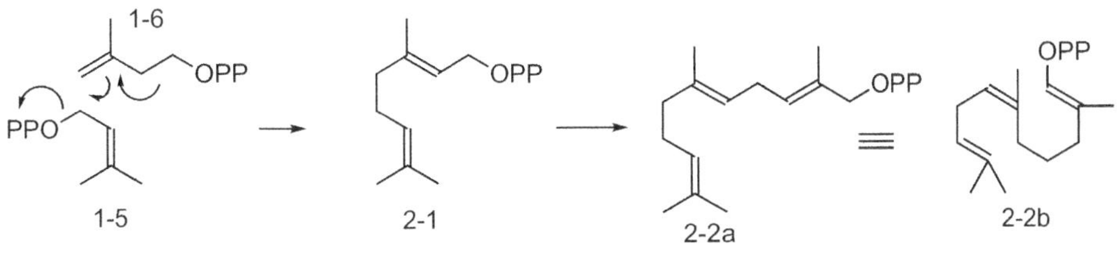

Scheme 2.2

The next step in the biosynthesis of steroids features an unusual head-to-head coupling reaction of two farnesol pyrophosphates (OPP, not shown in diagram) to afford the alicyclic triterpene squalene **3-1**, a compound found in shark liver oil (Scheme 2.3). Note that this product is in fact symmetrical about the newly formed bond. The next reaction in the sequence, which has only recently been uncovered, comprises oxidation of the terminal double bond to an epoxide. Opening of the oxirane leads to a domino-like series of ring-closing reactions and also concomitant migration of methyl groups. This chain reaction can be, and in fact has been, duplicated in the laboratory in the absence of enzymes. This series of reactions leads to the hypothetical steroidal carbocation **3-2**.

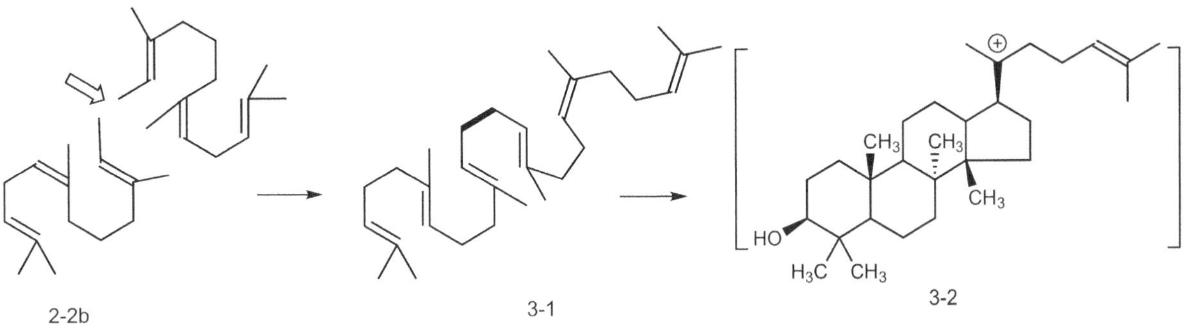

Scheme 2.3

The scheme for steroid biosynthesis is the same in both plants and animals up to the formation of the carbocation **3-2**. The biosynthesis diverges at this point: in animals the methyl group at C_8 migrates to afford lanosterol (**4-1**) as an isolable product (Scheme 2.4). The first steroidal product that can be isolated in plants, cycloartenol (**4-2**), features a cyclopropyl ring fused on to ring B at carbons 9,10.

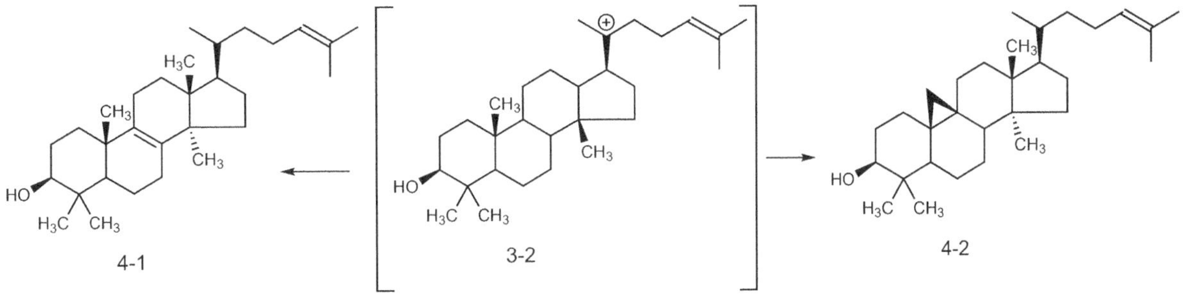

Scheme 2.4

A series of further transformations takes place in animals to transform C_{30} lanosterol to C_{27} cholesterol. In broad outline, the process involves first elimination of the methyl group located at C_{14} by an oxidative process to afford 14-desmethyllanosterol (**5-1**); a similar process then expels two methyl groups found at position 4 to give zymosterol (**5-2**) (Scheme 2.5). A series of enzyme-mediated steps in essence moves the double bond between rings B–C to C_5 and reduces the double bond in the side chain to afford cholesterol (**5-3**).

The side chain in cycloartenol (**4-2**) may undergo a series of further reactions in plants prior to conversion of the fused-ring nucleus to a steroid. Enzyme-catalyzed alkylation of cycloartenol can add a methylene group at side chain position 24 (**6-1**) (Scheme 2.6); oxidative elimination of the superfluous ring methyl groups in cycloartenol and opening

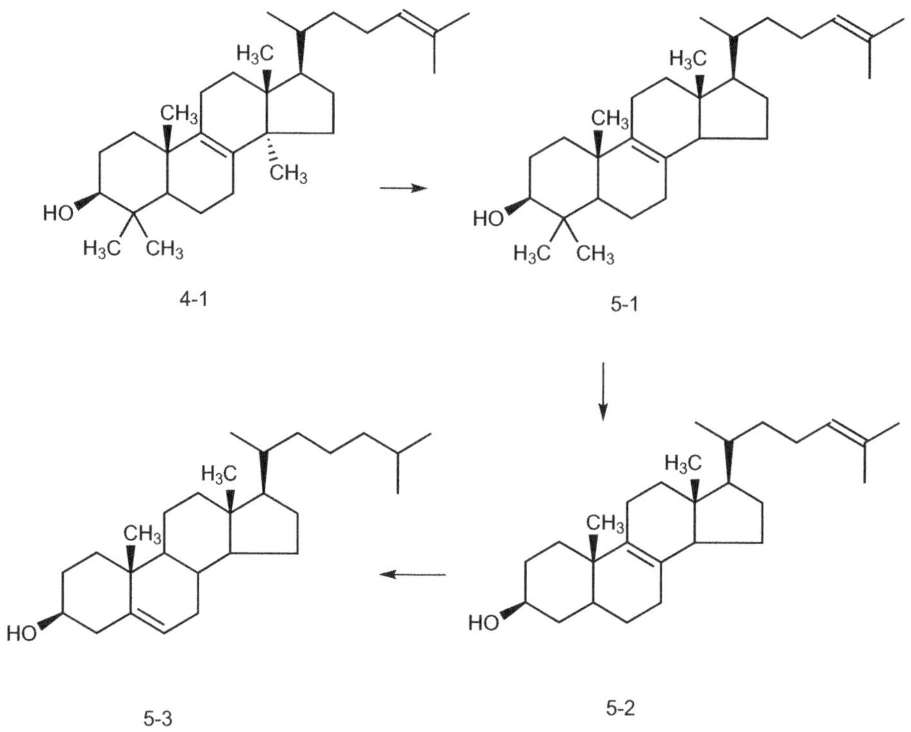

Scheme 2.5

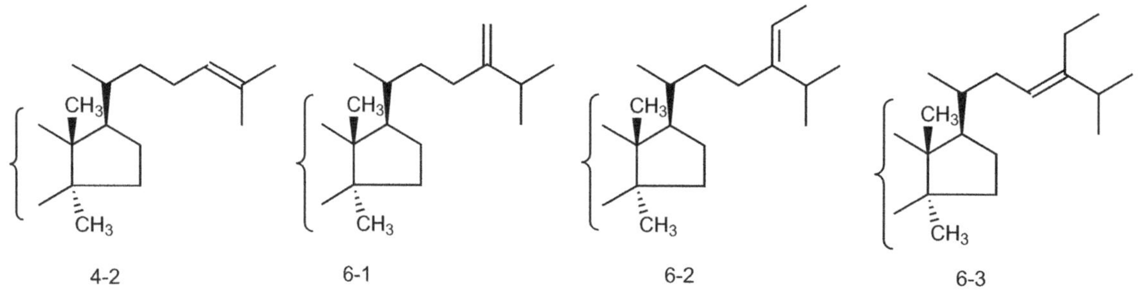

Scheme 2.6

of the fused three ring affords the plant sterol episterol. Further alkylation on the new side chain olefin gives intermediate **6-2**, which will proceed to the plant steroid avenasterol. Shifting the side chain double bond to positions 23–24 finally leads to **6-3**, an intermediate in the formation of stigmasterol, one of the important commercial steroid starting materials.

2.2 Commercial Steroid Starting Materials

Cholesterol and cholic acid comprise the most easily accessible source of bulk quantities of steroids. The lack of functionality in the side chain of cholesterol, however, stands in the way of shrinking it to the two-atom substituent at C_{17} that is present in progestins and corticosteroids. The relatively lengthy reaction sequences required for converting cholic acid to a suitable starting intermediate disqualified that source. Attention then turned to plant natural products for sources for steroid starting materials. The majority of plant steroids occur as saponins. These comprise compounds in which one or more carbohydrates are attached to steroids as glycosides. Those sugars are most often attached to the steroids via hydroxyl groups at positions 3 or at a side chain on ring D. The combination of the lipophilic steroid and hydrophilic sugar endows these derivatives with mild detergent-like properties. The soapy feel of saponin-containing plant sap is reflected by the name saponin (Latin *sapo*, soap). The steroid moiety in a saponin, variously called sapogenin or aglycone, can usually be liberated by relatively mild hydrolysis.

2.2.1 Diosgenin

The chemist Russel Marker conducted an intense search in the 1940s in Mexico for a steroidal natural product that could be used to prepare some of the hormonal steroids. His research soon centered on the Mexican wild yam *Dioscorea villosa* as a potential source of those compounds. Treating the saponin dioscin (**7-1**) with mild acid, he found, gave the aglycone diosgenin (**7-2**) (Scheme 2.7).

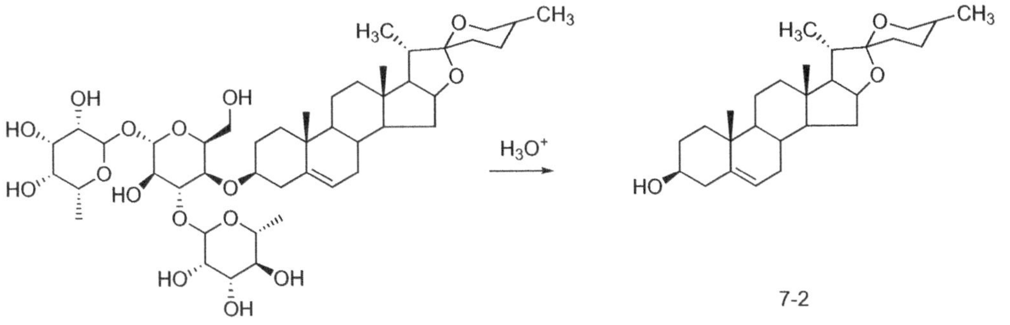

Scheme 2.7

Diosgenin is one of a sizeable group of naturally occurring steroids that feature a spiroacetal moiety. The structure of these sapogenins incorporates a reduced furan fused on to ring D with a perhydropyran attached to the furan (**7-2**). In terms of functionality, the moiety attached to ring D comprises an internal acetal of the carbonyl at C_{21} and hydroxyl groups at C_{16} and one at the end of the side chain (**8-1**) (Scheme 2.8). Treatment of diosgenin with acetic anhydride in all

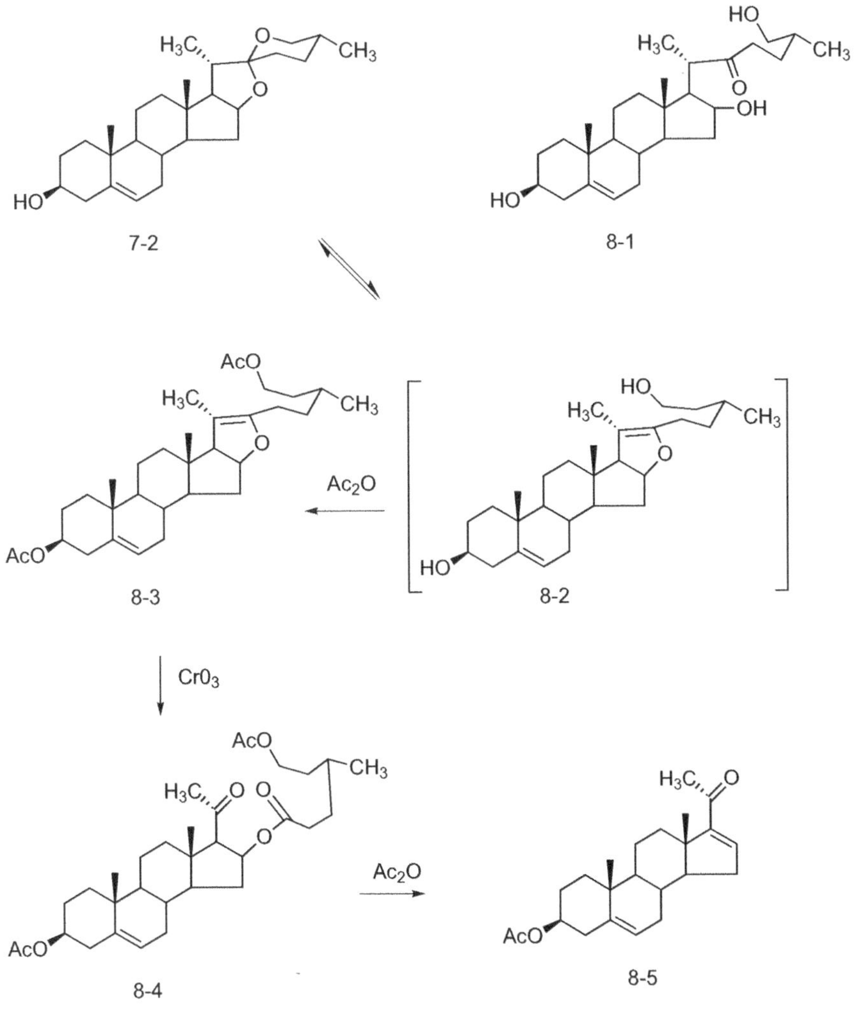

Scheme 2.8

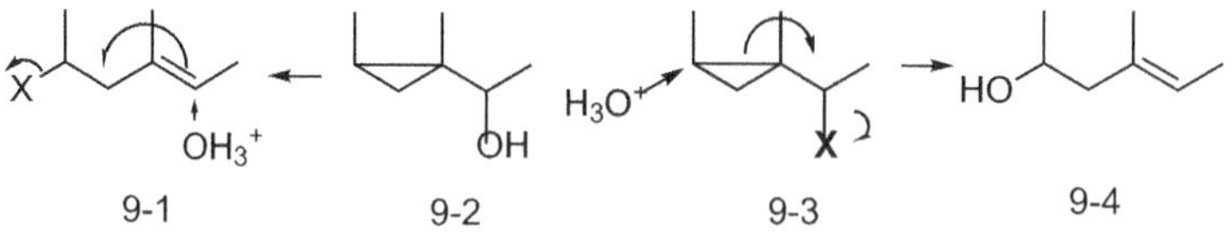

24 *Steroid Chemistry at a Glance*

Scheme 2.9

likelihood proceeds through the very small amount of the enol ether **8-2** that is formed by opening of the acetal. Acetylation of hydroxyl from the enol ether form locks that isomer in place; the hydroxyl at position 3 is also acylated in the process to afford **8-3**. Oxidation with chromium trioxide proceeds preferentially at the electron rich enol ether unsaturation to give **8-3**; this reaction in one fell swoop disposes of the superfluous side chain carbon atoms and introduces the carbonyl group at C_{20} present in many biologically active steroids. This last intermediate now comprises the ester of a β-hydroxyketone a species that readily undergoes β-elimination. Heating of **8-4** in acetic anhydride thus affords pregna-4,16-dien-3-ol-21-one (**8-5**). This last product now possesses the functionality at positions 3 and 20 required for preparing therapeutic steroids. The presence of the double bond in ring D offers the possibility of introducing substituents at an 'unnatural' position. Transformations of this now readily available chemical to commercially important steroids will be found throughout the following chapters.

An approach to modifying the structure of rings A and B prior to stripping off the spiroacetal relies on the homoallyl–cyclopropyl rearrangement. This reaction was actually discovered in work on steroids and for many years bore the title *i*-steroid rearrangement. This reaction involves addition of a reagent to the olefin of a homoallylic system that bears a good leaving group two atoms removed from the other end of the double bond (**9-1**) (Scheme 2.9). The reaction proceeds to formation of a cyclopropane with concomitant expulsion of the leaving group **X** (**9-2**). The system can be reversed by providing a good leaving group on the methylene group adjacent to the cyclopropyl moiety (**9-3**). In this case, attack by the reagent on one end of the cyclopropyl opens the ring an expels the leaving group **X**. This reaction

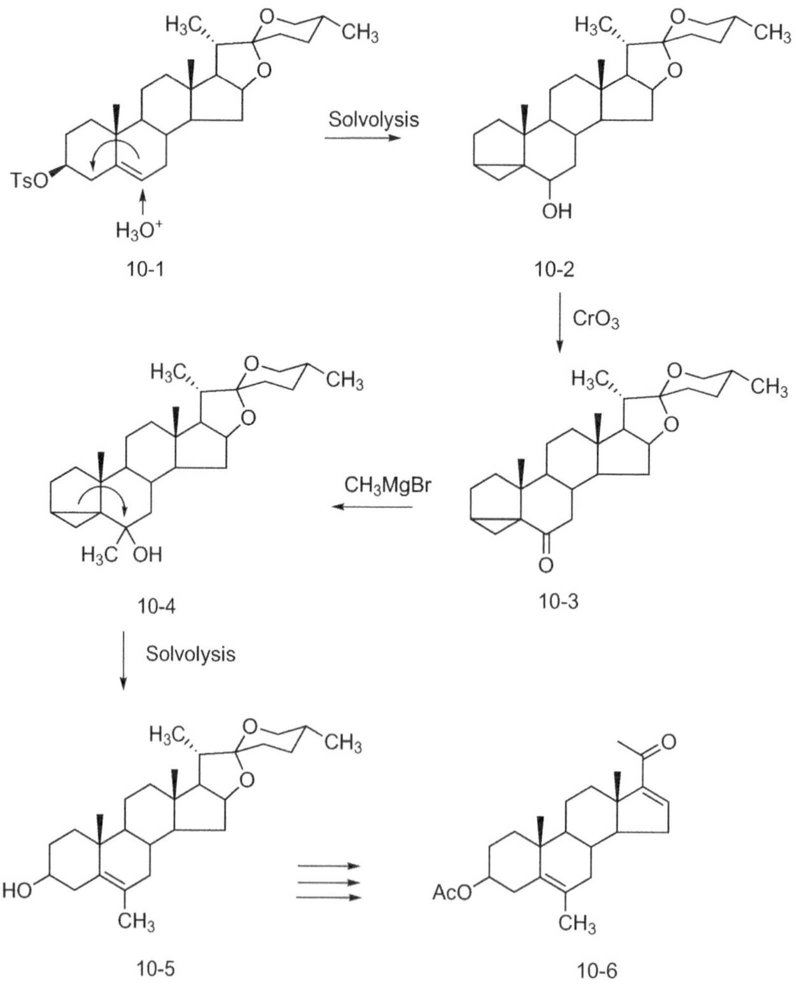

Scheme 2.10

is particularly favored in systems in which the reacting centers are rigidly constrained, as is the case in steroids, to resemble the hypothetical transition state.

In the system at hand, diosgenin (**7-2**) is first converted to its 3-toluenesulfonate **10-1** by reaction with *p*-toluenesulfonyl chloride (Scheme 2.10). Solvolysis of this compound under weakly acidic conditions leads to displacement of the excellent leaving group *p*-toluenesulfonate and formation of the cyclopropane-containing derivative **10-2**. The newly formed hydroxyl group is next oxidized with chromium trioxide to give the 6-ketone **10-3**, reaction of which with methylmagnesium bromide gives the carbinol **10-4**. The thus -formed tertiary alcohol is particularly sensitive to displacement. Solvolysis of **10-4** leads to reversal of the rearrangement and formation of the 6 methyl analogue **10-5** of diosgenin (**7-2**). The methylated analogue is next subjected to the series of reactions that lead to the pregnenolone to afford the 6-methyl analogue **10-6**.

2.2.2 Soybean Sterols

An adventitious finding led to what is arguably the most important source for bulk steroid starting materials. Investigating the nature of a white solids that had precipitated in a tank of soybean oil, the chemist Percy Julian identified them as a mixture of plant sapogenins that consisted mainly of a mixture of stigmasterol (**11-1**) and its side chain reduction product sitosterol (**11-2**) (Scheme 2.11). The presence of the precipitate was attributed to hydrolysis of plant saponins caused by moisture that had made its way into the tank. Specific solvent combinations have been developed for separating stigmasterol, which has a double bond in the side chain from sitosterol, which lacks such a potential point of attack.

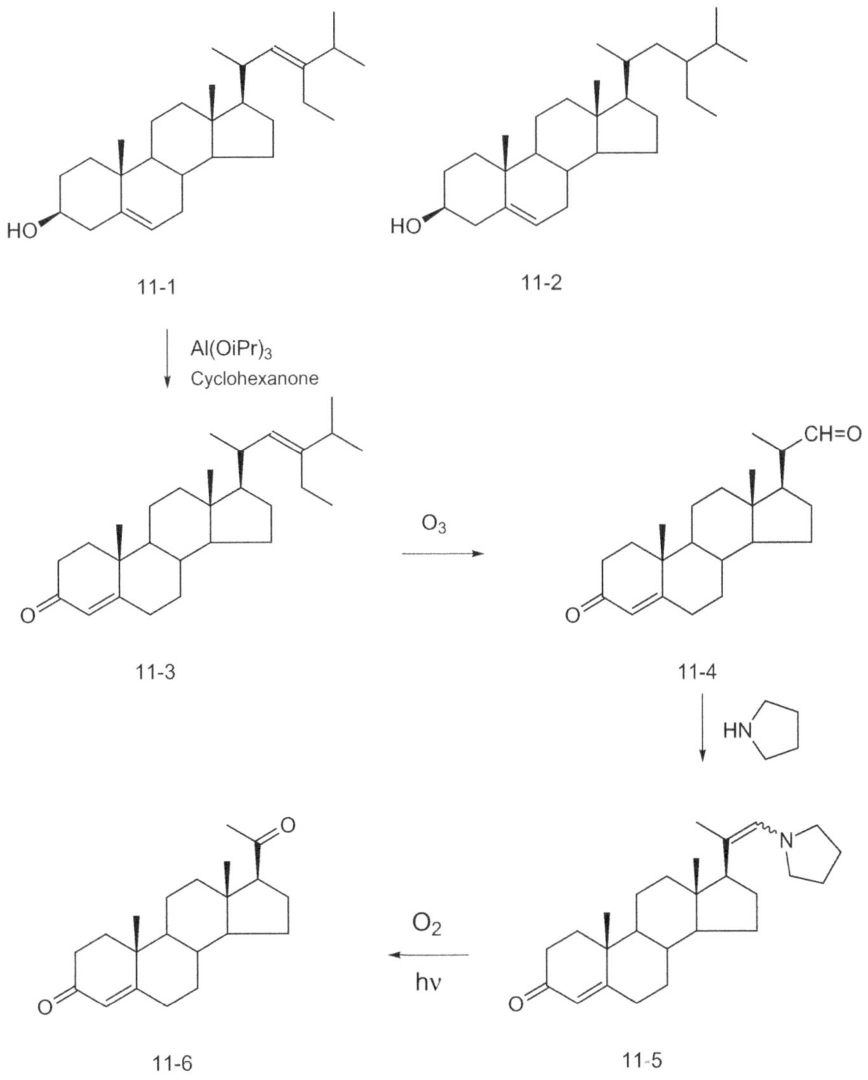

Scheme 2.11

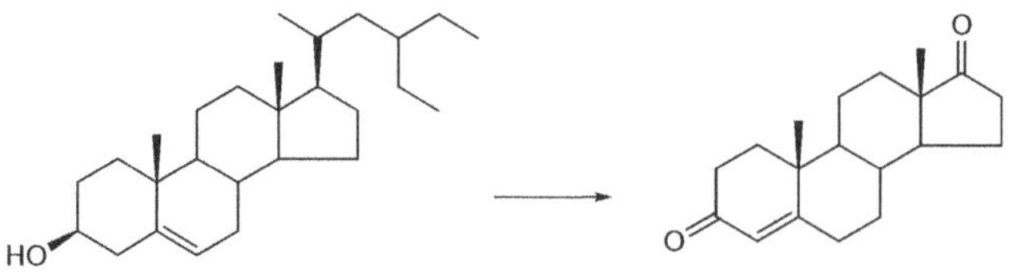

Scheme 2.12

The first problem that needs to be addressed lies in differentiating the chemical reactivity of the side chain double bond from that at positions 5 and 6 in ring B. The Oppenauer reaction comprises one of the standard operations in steroid chemistry. This very mild reaction in essence involves the transfer of a pair of hydrogen atoms from the steroidal carbinol to a ketone such as acetone or cyclohexanone added in the reaction solvent. A metal alkoxide, commonly aluminum, acts as the transfer agent. In the case at hand, reaction of stigmasterol, **11-1** with cyclohexanone in the presence of aluminum isopropoxide leads to the 3-keto derivative. The double bond in the starting material then shifts into conjugation under the slightly basic reaction conditions to afford the corresponding 4-en-3-one **11-3**. Reaction of that intermediate with ozone proceeds preferentially at the unsaturation in the side chain due to the higher electron density at that site compared with that of the enone. Workup of the ozonide then affords aldehyde **11-4** in which the larger part of the side has been clipped off. Removal of the last superfluous carbon atom begins with treatment of the intermediate **11-4** with pyrrolidine. This amine again reacts preferentially with the aldehyde group rather than the less reactive conjugated ketone at position 3 to yield the enamine **11-5**. This reaction moves one end of the double bond to position 20. Photo-oxidation of **11-5** cleaves that double bond, in essence clipping the last superfluous carbon atom while moving the carbonyl group to the now two-atom side chain. The product, progesterone (**11-6**), is a drug in its own right in addition to being a key starting material for many other steroids.

The at one time superfluous sitosterol **11-2** has found minor use as a drug for treating elevated cholesterol levels by inhibiting the absorption of dietary cholesterol. Methods have been developed within the past decade for converting sitosterol to androst-4-ene-17,20-dione (**12-1**) by fermentation (Scheme 2.12).

3
Estranes: Steroids in Which Ring A is Aromatic

3.1 Biological Activity

The apparent structural simplicity of the estrane nucleus belies the importance of this class of steroids in animal species. In women, blood levels of estradiol (**1-1**) and estrone (**1-2**) and related estranes (Scheme 3.1) rise and fall every month over the four or more decades from late puberty to menopause. These hormonal steroids and their C_{21} companion, progesterone (**1-3**), which exhibits analogous cyclic changes in blood levels, regulate the reproductive cycle and the subsequent viability of a fertilized ovum. In addition to their role in the reproductive cycle, estrogens are directly involved in the maintenance of many female gender-related structures such as the reproductive organs and breasts. This activity reflects the regulation of protein synthesis by steroids by way of direct interaction with DNA. Estradiol and some of its esters, in addition to estrone itself, have been used to treat estrogen deficiency syndromes from the time when commercial amounts of those steroids became available in the 1930s. Estrogens have subsequently found use in treating adverse effect of menopause. A more important application resides in the circumstance that estrogens have formed part of oral contraceptives starting with the initial approval of 'The Pill' in the early 1960s.

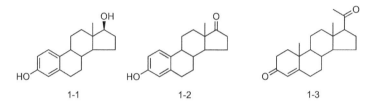

Scheme 3.1

3.2 Sources of Estranes

There exists a plethora of plant natural products built upon the steroid nucleus. Virtually none of those phytosteroids include an aromatic ring A in their structure. Much the same holds for the animal kingdom, with an important exception: sizeable amounts of estrone-related estrogens are present as their sulfated derivatives in the urine of pregnant mares. Although the amount of estrogens in the urine is too small to be considered a source for compounds for further chemical modification, the crude mixture of sulfated estrogens (Premarin®) derived from pregnant mare's urine has been used as an estrogen supplement for many years.

3.2.1 From Androstanes
3.2.1.1 From 1,4-Dien-4-ones

One of the more readily available plant steroids, dehydroepiandrosterone (DHEA), provides the raw material for conversion to estrone. An early step in converting steroids from plant sources to starting material for modified estrogens comprises obligate expulsion of the methyl group at position 10 that blocks aromatization of the A-ring.

The dienone **2-6** is the initial target in a widely used scheme for aromatizing ring A (Scheme 3.2). It noteworthy that this ring, with the exception of the methyl group at C_{10}, is already in the same oxidation state as the desired phenol. The sequence for the conversion to estrone starts with the Oppenauer oxidation of DHEA to androstene-3,17-dione (**2-2**). The double bond shifts into conjugation in the course of the reaction. This migration, as will be seen in the pages that follow, occurs in many other oxidations of 3-keto-5-ene steroids. Catalytic reduction of **2-2** adds hydrogen from the less hindered backside to afford the 5α isomer **2-3**. Treating a solution of the latter in acetic acid with bromine proceeds initially to the 4,4-dibromide (**2-4**). This intermediate, which is not isolated, rearranges spontaneously to the 2α,4α-dibromide in which each of the halogen atoms is actually equatorial. Treatment of **2-5** with a non-nucleophilic base such as 2,4,6-collidine leads to double dehydrobromination and thus formation of the 2,4-diene, **2-6**. In early work, a solution of this dienone in mineral oil was passed through a column of glass beads heated to 600°C, a temperature that cracked some of the mineral oil. This resulted in expulsion of a methyl group and consequent conversion of ring A to a fused benzene.

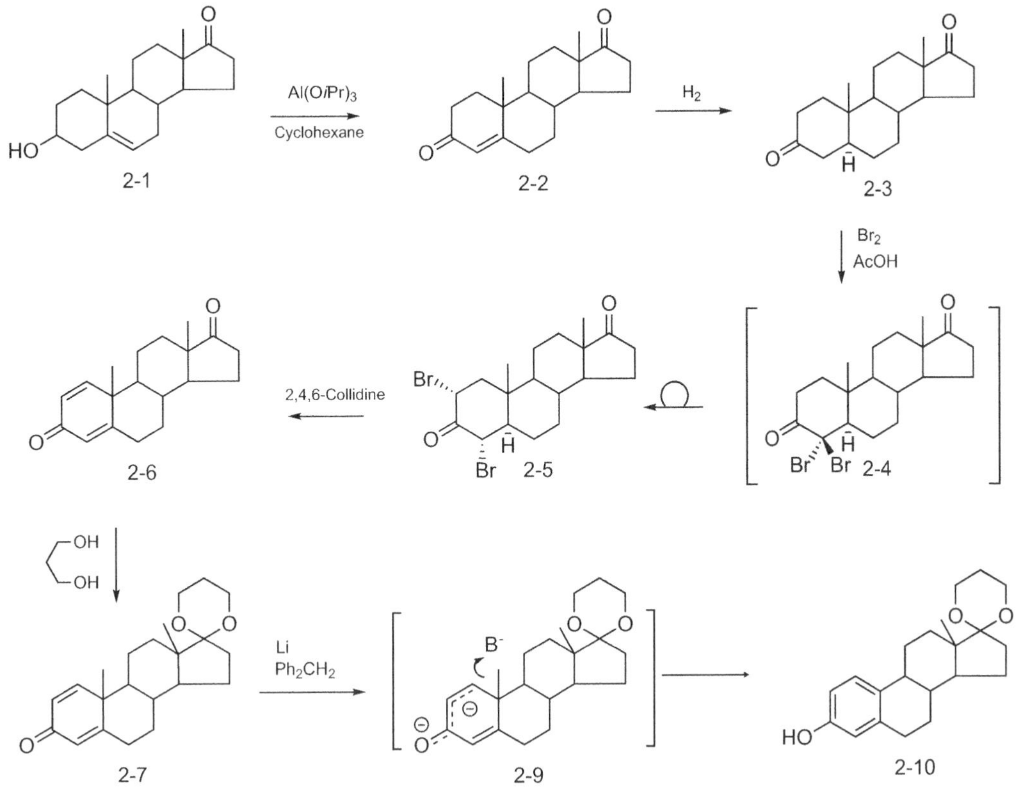

Scheme 3.2

A marked improvement on what amounted to pyrolysis consists in treatment of the dienone with strong base. In the case at hand, the carbonyl group at position 17 needs to be protected against attack by the lithio reagent. The ketone at position 17 is therefore first converted to its trimethylenedioxy acetal **2-7** by reaction with propylene-1,3-diol in the presence of a small amount of acid. Treatment of this 1,4-dien-3-one with lithium metal in the presence of diphenylmethane leads to the trimethylenedioxy acetal, **2-9**, of estrone. The diphenylmethane present in the reaction mixture presumably quenches the extruded methyllithium to prevent its addition to the starting dienone. Treatment with dilute acid then restores the carbonyl group to afford estrone (**1-2**).

The venerable dienone–phenol rearrangement (**3-1** → **3-2**) offers another way for preparing steroids with an aromatic A-ring (Scheme 3.3). The simplest case of this method illustrates a serious drawback to this approach: one of the geminal methyl groups shifts onto the adjacent position.

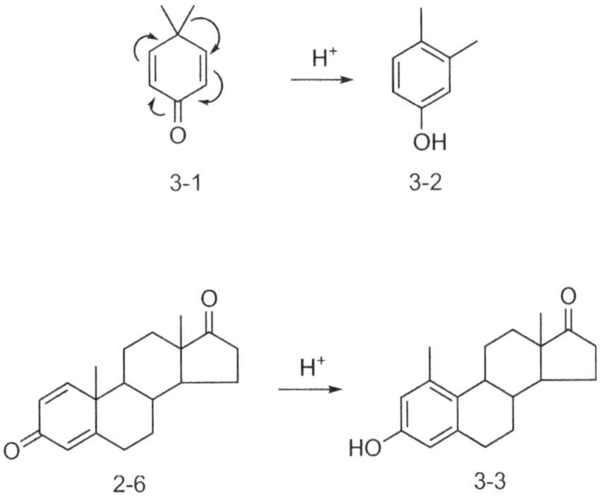

Scheme 3.3

In the case of the steroid dienone **2-6**, the methyl group formerly on C_{10} migrates to position 1 on the aromatic ring (**3-3**). This compound and its derivatives show much of the same biological activities as the corresponding compounds lacking the methyl group.

3.2.1.2 Functionalizing the Angular (C_{19}) Methyl Group

Ring-A-aromatic steroids are formed biologically from androst-4-ene-3,17-dione. A series of enzymes, collectively named aromatases, mediate the stepwise oxidation of the methyl group at position 10 to carboxaldehyde (**4-1**) (Scheme 3.4). The functionality in ring A in essence consists of a vinylogous β-dicarbonyl function, an array known readily to lose one of the carbonyl carbons. The enzyme elimination aromatase then catalyzes expulsion of the angular carbonyl function to afford estrone (**1-2**).

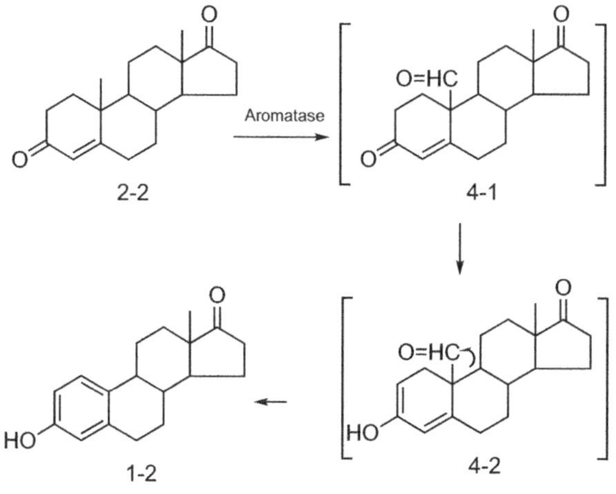

Scheme 3.4

Several methods have been developed that to some extent mimic the aromatase enzyme. The schemes share the goal of transforming the methyl group at position 10 into a function that can be more readily expelled from substituted A-rings. Note, however, that these procedures are more relevant to the preparation of gonanes than estranes (see Chapter 4). The chemistry for functionalizing the angular carbon on position 10 depends on the close proximity in space of an ester of a 6β-hydroxyl group to C_{19} (see **5-1**, Scheme 3.5). One example starts with esterification of 6β-hydroxy-β-dihydrocholesteryl acetate with nitrous acid or, alternatively, nitrosyl chloride. Irradiation with a mercury vapor lamp of a solution of the nitrite ester (**5-2**) in toluene leads to transfer of the nitrite across space to the angular methyl group. The first-formed nitrite dimer is then heated to afford the 19-oxime (**5-3**) The newly formed oxime can then be converted to an aldehyde by treatment with nitrous acid.

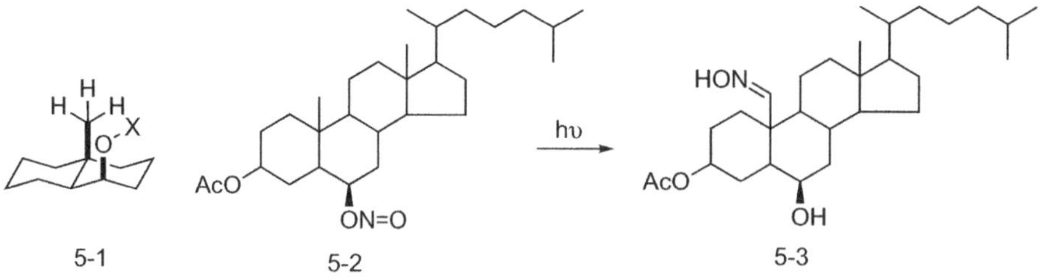

Scheme 3.5

Essentially the same transformation can be conducted on a more highly substituted steroid. The required nitroso ester can in principle be prepared from reaction of DHEA 3-acetate (**6-1**) with bromine and nitrous acid (Scheme 3.6). The reaction sequence would start by addition of bromine from the open α-side of the steroid to form bromonium ion **6-2**. Attack on that ion by the nitroso anion from the β side together with the rule of diaxial opening of a three-membered ring will the lead to formation of the nitrite **6-3**. Irradiation of a toluene solution of the nitrite ester with a mercury vapor lamp walks the nitro group across the gap, in effect functionalizing C_{19}. Heating the initially formed nitroso dimer in propan-2-ol yields the 19-oxime **6-4**. This too affords the aldehyde on reaction with nitrous acid.

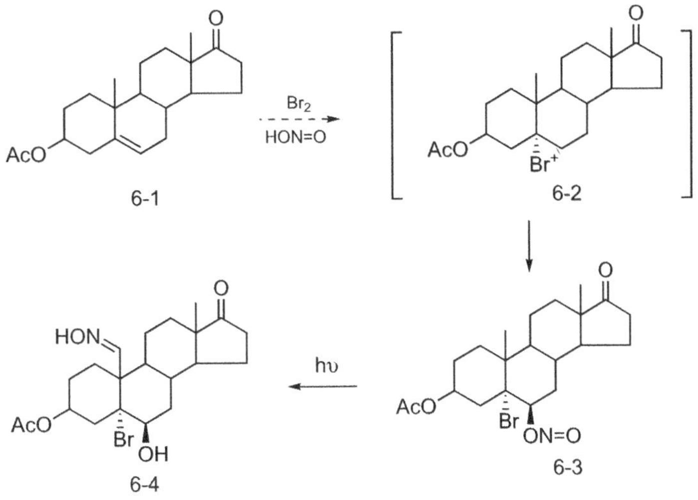

Estranes: Steroids in Which Ring A is Aromatic   31

Scheme 3.6

Direct introduction of oxygen on to the angular methyl group requires the presence of an oxidizing agent such as lead tetraacetate. The scheme in a typical example starts with the addition of hypobromous acid to the unsaturation in pregnenolone acetate (**7-1**) (Scheme 3.7). Formation of the bromohydrin **7-2** follows the same course as that described above (**6-1** → **6-3**). Reaction of the bromohydrin with just over an equimolar amount of lead tetraacetate gives the cyclic ether **7-3** from attack on the angular methyl group at position 10. Treatment of **7-3** with boron trifluoride in the presence of acetic anhydride opens the newly formed tetrahydrofuran to the 6,19 diol. This then reacts with acetic anhydride to give the corresponding acetate, **7-4**. Reaction of **7-2** with chromium trioxide leads to oxidation of the only open methylene group to a carbonyl function, forming what is now a lactone (**7-5**). Alkaline hydrolysis opens the lactone to afford the carboxylic acid, **7-6**.

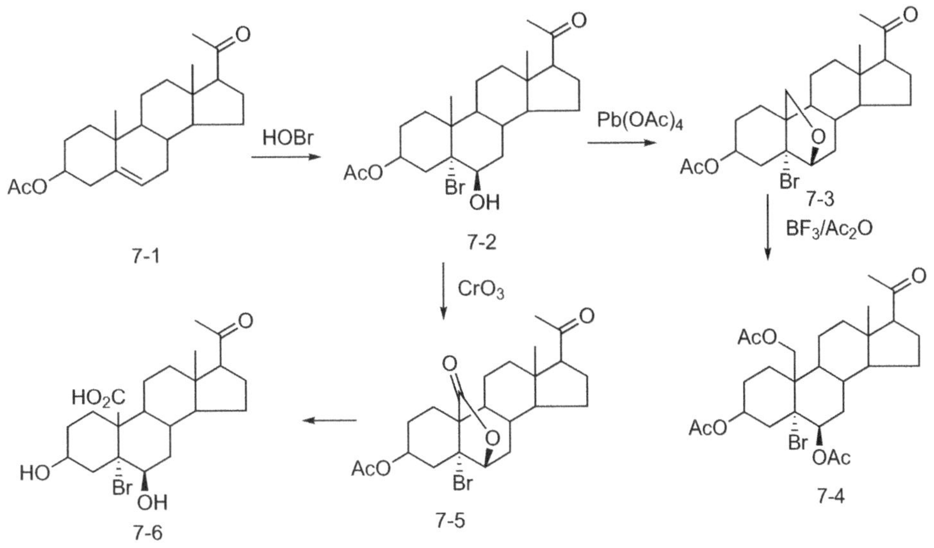

Scheme 3.7

An interesting application of this remote oxidation reaction invokes the homoallyl–cyclosteroid rearrangement (see Chapter 2, Scheme 2.9). Solvolysis of the tosylate from cholesterol (**8-1**) gives the corresponding cyclosteroid, **8-2** (Scheme 3.8). Heating a solution of that product with lead tetraacetate leads as above to attack on the C_{19} angular methyl group, thus affording a moderate yield of the tetrahydrofuran **8-3**. The newly formed tetrahydrofuran ring is then opened, for example with boron trifluoride. The two hydroxyl groups in the resulting diol, **8-4**, are then converted to their tosylates **8-5** by means of toluenesulfonyl chloride. Solvolysis of this intermediate reverses the rearrangement, restoring the 3-hydroxy-5-ene array (**8-6**). This intermediate can then in principle be elaborated to a compound in which the hydroxymethyl substituent becomes a good leaving group.

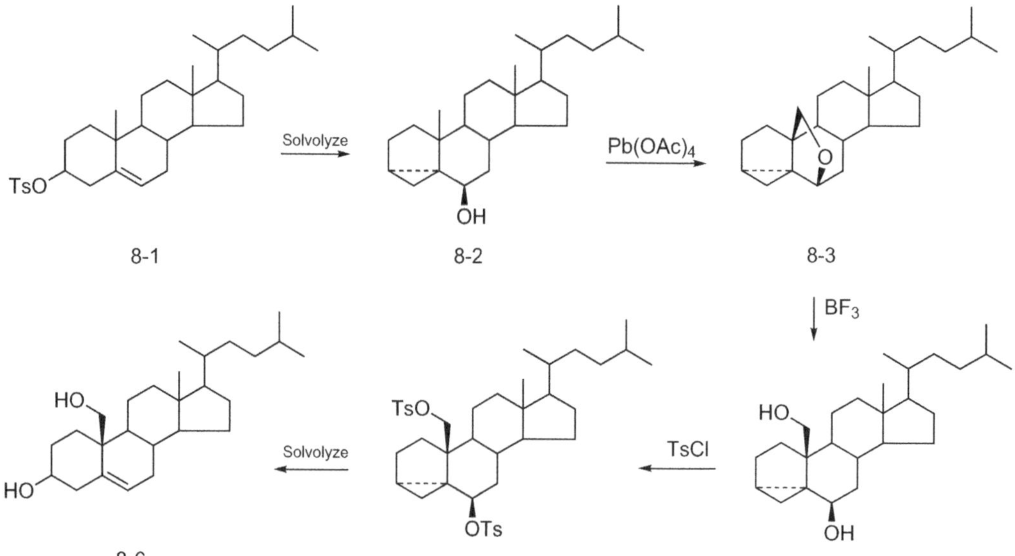

3.2.2 Estrogens by Total Synthesis

The relative simplicity of the structure of estranes compared with those of other classes of steroids has made this structural class an attractive target for work aimed at total synthesis from so-called coal tar starting materials. About half a dozen syntheses have been developed, one of which, the Smith–Torgov synthesis, is probably still used commercially.

3.2.2.1 W. S. Johnson Synthesis

Minor amounts of the further dehydrogenated estranes equilin (**9-1**) and equilenin (**9-2**) are to be found in pregnant mare's urine (Scheme 3.9). The chiral centers at C_8 and C_9, it should be noted, have been replaced by trigonal carbon atoms devoid of chirality in equilenin. The reduced number of possible enantiomers of **9-2** led to the selection of this compound as the target for an early total synthesis.

A synthesis developed by the group led by W. S. Johnson starts with the cyclization of the naphthylbutyric acid **10-1** catalyzed by a strong acid such a liquid hydrogen fluoride (Scheme 3.10). Condensation of the product **10-2** with ethyl formate and sodium ethoxide gives the corresponding formylated derivative **10-3**, depicted in Scheme 3.10 as its enol tautomer. Reaction of the formyl derivative **10-3** with hydroxylamine in acetic acid likely proceeds to give initially the oxime of the formyl group. That then cyclizes to the isoxazole **10-4** under the reaction conditions. Treatment of the isoxazole with strong base such as potassium *tert*-butoxide then leads to the ambident anion **10-5**.

The methyl group intended for position 18 in equilenin is then introduced by reacting the ambident anion **10-5** with methyl iodide. Stobbe-like condensation of the product **11-1** with methyl succinate and *tert*-butoxide can in theory start by addition of the succinate anion to either the ring carbonyl group or the nitrile. Assuming that the reaction follows the latter course it will initially proceed to give an imine such as **11-2**. The remaining excess base then again abstracts a proton from the succinate. This adds to the ring-ketone to form the five-membered D-ring. Work-up affords intermediate **11-3**. A series of routine steps that include reduction of the unsaturation in ring D and subsequent removal of the superfluous carbomethoxy function finally afford equilenin (**9-2**).

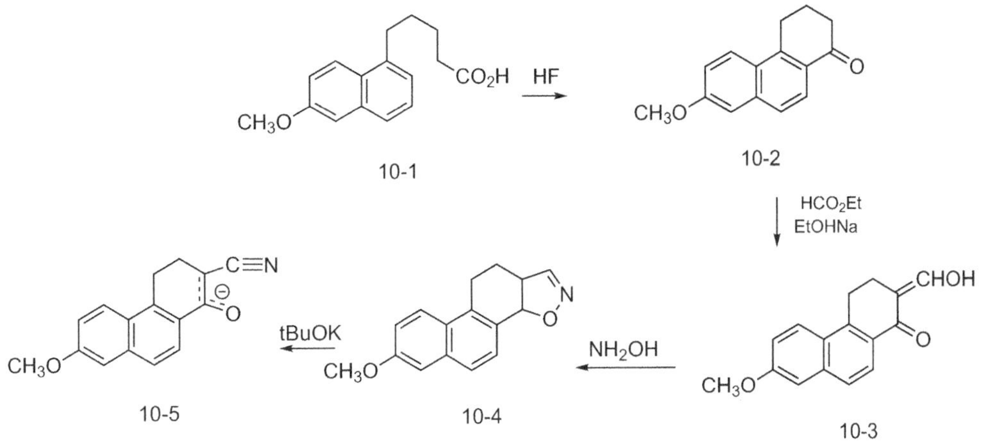

Scheme 3.10

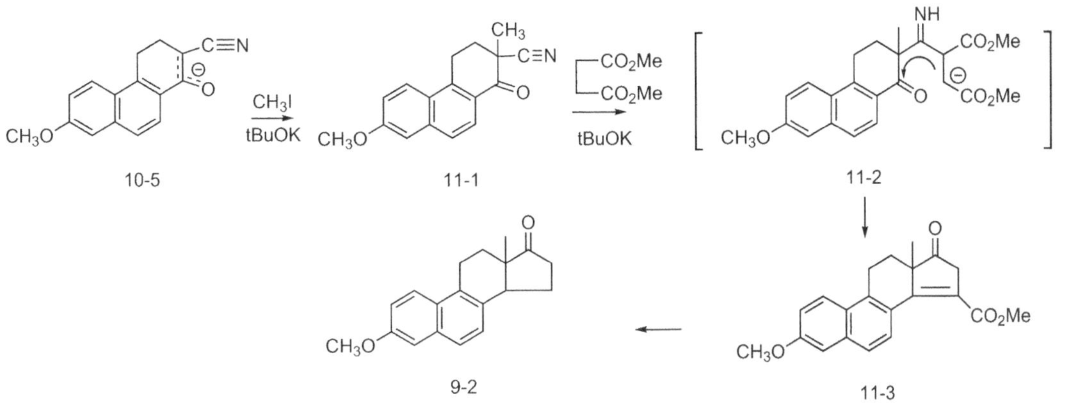

Scheme 3.11

3.2.2.2 Anner and Miescher Synthesis

The first total synthesis of estrone itself by Anner and Miescher in 1948 also ended with the construction of the five-membered ring D. This synthesis differs from that above mainly by the replacement of the naphthalene moiety by a tetralin. Dieckmann condensation of the ester **12-1** gives the hydrophenanthrone **12-2** (Scheme 3.12). Treatment of the

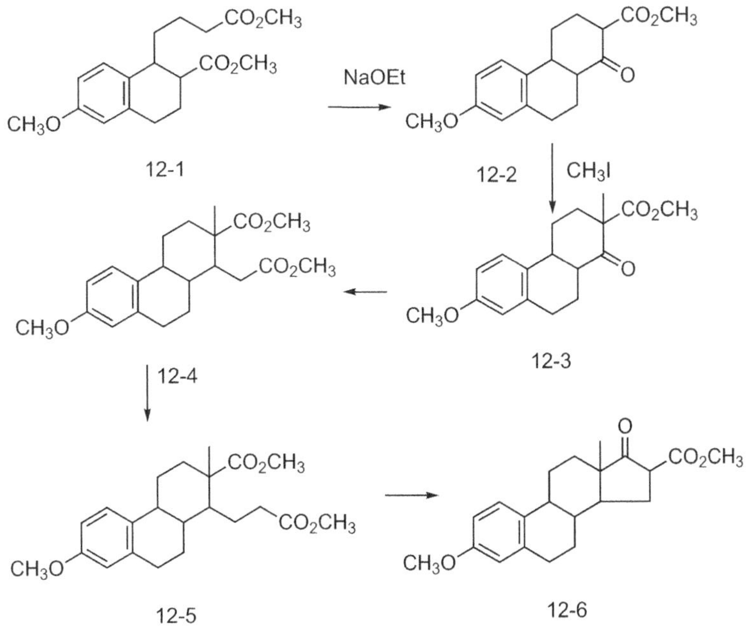

Scheme 3.12

anion from that product with methyl iodide introduces the future angular methyl group (**12-3**). Two of the remaining carbon atoms for ring D are then added by means of a Favorskii reaction (methyl chloroacetate and zinc).

Dehydration of the thus-produced tertiary hydroxyl group introduces unsaturation at positions 14–15 (steroid numbering). Catalytic hydrogenation of that product then affords keto ester **12-4** as a mixture of diastereomers. These are separated, and the longer side chain in the isomer with the proper stereochemistry is then extended by one carbon atom, to afford **12-5**. A second Dieckmann cyclization gives the keto ester **12-6**. The product from that reaction is then saponified and the resulting keto acid (**13-1**) allowed to decarboxylate to afford estrone methyl ether (**13-2**) (Scheme 3.13). Cleavage of the methyl ether in ring A then affords estrone (**1-2**).

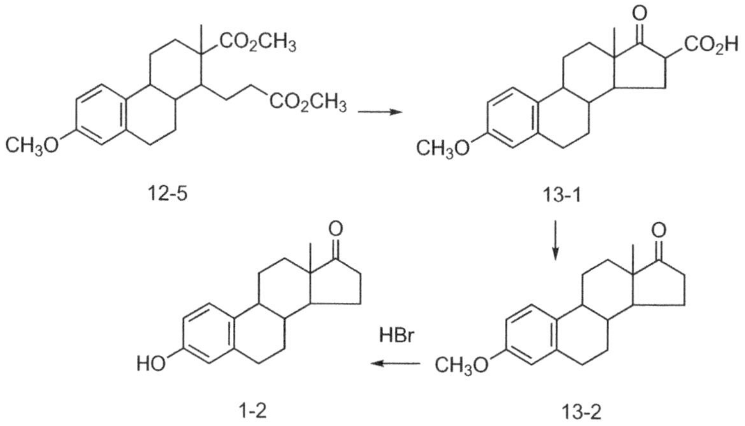

Scheme 3.13

3.2.2.3 Smith–Torgov Synthesis

The scheme that is likely still used to prepare derivatives of estrone and also some modified estranes on a commercial scale was developed independently by Smith in the USA and Torgov in the former USSR.

This relatively concise synthesis starts by addition of the Grignard reagent from vinyl bromide to the methoxytetralone **14-1** (Scheme 3.14). The key step comprises condensation of that vinyl alcohol with 2-methylcyclopenta-1,3-dione (**14-4**) in the presence of a catalytic amount of base. The transformation can be envisaged as involving initial dehydration of the tertiary carbinol in **14-2** prompted by the surprisingly acidic dione to afford the ambident olefin **14-3**. The enolate from **14-4** then adds to one end of that olefin. That transient product then closes to form a six-membered ring, to afford the tetracyclic product, **14-5**, in effect a steroid nucleus. Catalytic hydrogenation proceeds to add hydrogen to the 14–15 olefin from the more accessible α side. This locks the C–D ring fusion into the disfavored *trans* configuration. The carbonyl group at position 17 is then reduced to the corresponding carbinol by means of sodium borohydride. This also involves addition of a proton from the backside to give the 17β-hydroxyl. Reduction the

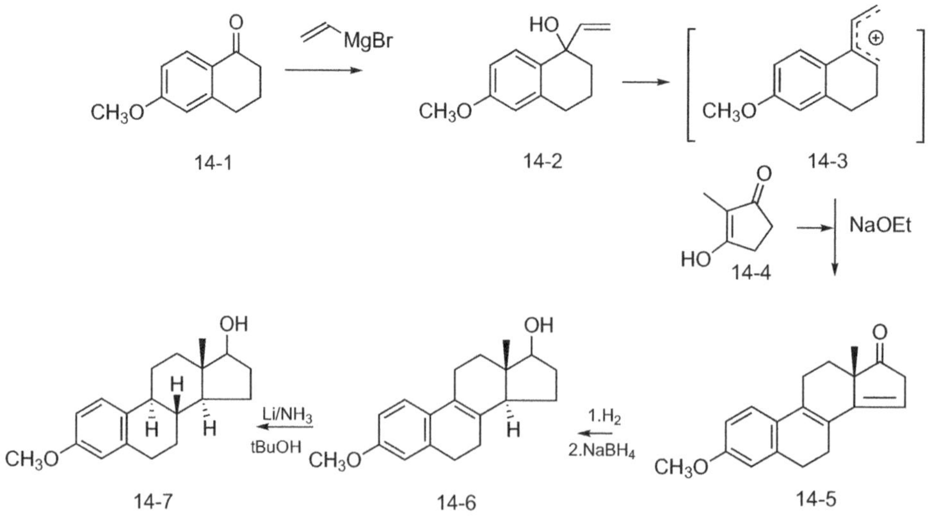

Scheme 3.14

remaining, styrene-like, double bond in product **14-6** is accomplished by treatment with lithium, *tert*-butanol and liquid ammonia (this reaction, the Birch reduction, will be occur with some frequency in Chapter 4). This reaction proceeds to form a product with the *trans* B–C ring fusion and thus estradiol 3-methyl ether (**14-7**). More recently, it has been reported that the stereochemical outcome of this scheme relies heavily on the location of the olefin in ring D at positions 14–15.

In that example, steric pressure due to the presence of two methyl groups at C_6 in tetralol **15-1** causes the unsaturation in ring D of the condensation to migrate to position 15–16 (Scheme 3.15). At the same time, the C–D ring junction assumes the favored *cis* configuration, bending ring D away from the geminal dimethyl grouping. The stereochemistry of the subsequent Birch reduction also goes awry, yielding product **15-4** in which the B–C ring fusion also assumes the *cis* configuration. The overall result is a product that looks like a steroid only on a flat piece of paper.

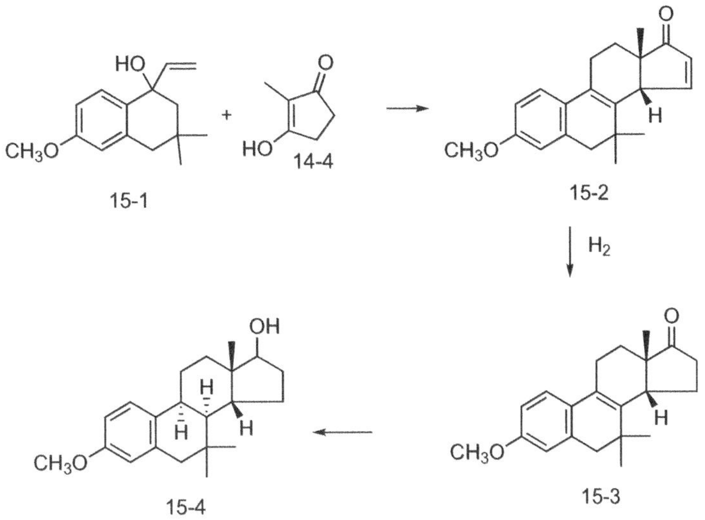

Scheme 3.15

3.2.2.4 Corey Route

The preceding schemes for preparing estrone have the disadvantage that they yield the steroid as a mixture of the two enantiomers. Preparing material identical with that which occurs in Nature requires resolution of those diastereomers. This also implies loss of half of the mass of final product. A synthesis that produces estrone and its derivatives directly without the need for that extra step depends on the use of chiral auxiliaries in the formation of ring C. The crucial step in this synthesis involves Diels–Alder condensation of the diene **16-1** with the fumaric ester aldehyde **16-2** in the presence of the oxazaborolidinium salt shown in Figure 3.1; this reaction affords the tricyclic intermediate **16-3** as a single enantiomer (Scheme 3.16).

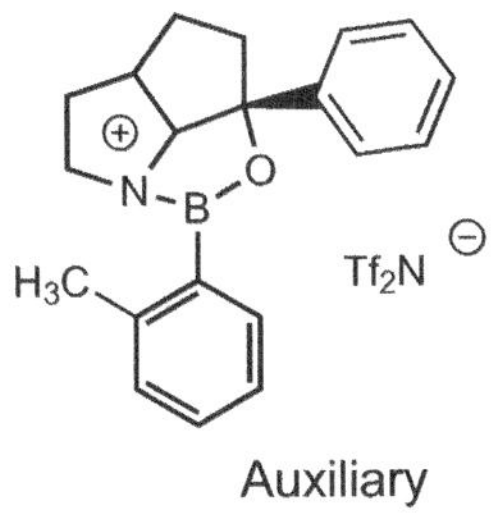

Auxiliary

Figure 3.1 Auxiliary oxazaborolidinium salt

The additional carbon atom required for building the five-membered ring D is then added by means of methylmagnesium bromide (**16-4**). The ester at future C_{15} is reduced to a carbinol by reaction with lithium aluminum hydride. Swern oxidation (dimethyl sulfoxide and oxalyl chloride) next serves to oxidize each of the two resulting hydroxyl groups to carbonyl functions (**16-5**). Dieckmann-like base-catalyzed condensation then closes the five-membered ring D. Catalytic hydrogenation in this case reduces the extended polyene in a single step to afford estrone 3-methyl ether (**13-2**) as a single diastereomer.

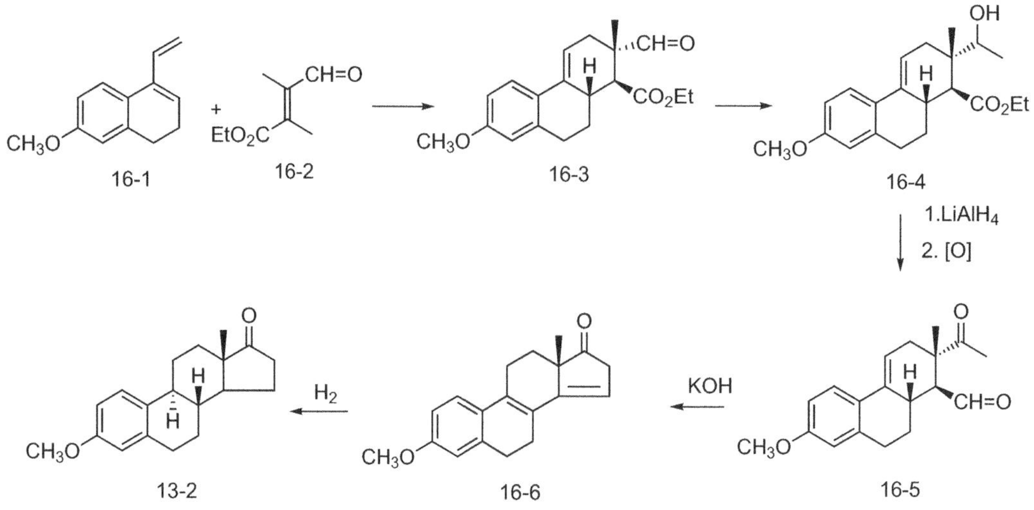

3.2.2.5 Pattenden Synthesis

The complex anisole derivative **17-1** comprises a late intermediate in a synthesis of estrone that culminates with a cascade of free radical cyclizations. Condensation of the carbonyl group in that compound with the phosphorane from diphenylmethoxymethylenephosphine gives the aldehyde acetal **17-2** (Scheme 3.17). The silyl protecting group on the other chain is then removed by exposure to fluoride ion. The thus-revealed alcohol is converted to an iodo group by a Mitsonobu reaction with iodine. Reaction of this intermediate (**17-3**) with the classical free radical initiators dibutyltin hydride and azobisisobutyronitrile (AIBN) generates a transient free radical at the site of the halogen. That reactive species then undergoes radical cyclizations depicted in **17-4** to afford the steroid **17-5**. Oxidation of the methyl ether at C_{17} with chromium trioxide affords estrone 3-methyl ether as a mixture of its two diastereomers.

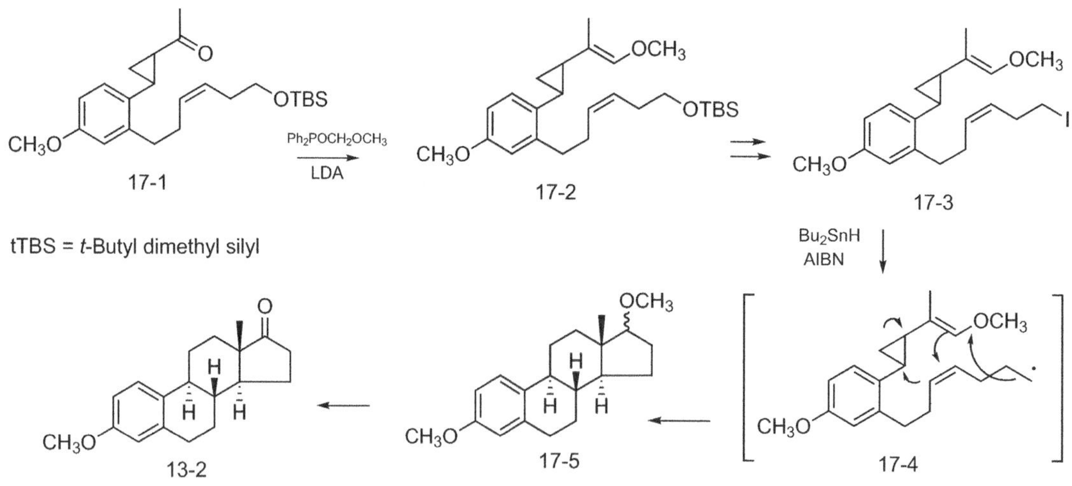

3.2.2.6 Synthesis of 2-Methoxyestrone

Reactions aimed at introducing additional substituents on the aromatic ring of estranes via standard methods commonly lack regiospecificity (see Section 3.3). A synthesis targeted at preparing estrone with additional oxygen at position 2 in effect comprises yet another total synthesis. The general approach, building rings B and A on to a unit that represents rings C–D, is conceptually close to that which has been used to prepare gonanes (see Scheme 4.3, Chapter 4).

Reaction of hydrindanone **18-1** with Grignard reagent **18-2** leads to selective reaction of a single organometallic reagent on the lactone carbonyl group (Scheme 3.18). The selectivity for the lactone is presumably due to the fact that the cyclopentanone carbonyl group is apparently sufficiently less reactive due to steric hindrance. Oxidation, under basic conditions, of the newly revealed hydroxyl in the cyclohexane ring leads to the diketone **18-4**. Base then catalyzes the ring-forming aldol reaction to yield the cyclohexenone **18-5**. Mild acid treatment leads to hydrolysis of the acetal group (**18-6**); reaction of the new diketone with base leads a second cyclization, affording the steroid nucleus (**18-7**). The oxidation state of ring A now corresponds to a phenol. Mild acid leads to reorganization of the bonds in ring A to form an aromatic ring and thus 2-methoxyestrone (**18-8**). Sodium borohydride reduces the ketone at position 17 to the β-alcohol, yielding 2-methoxyestra-3,17β-diol (**18-9**).

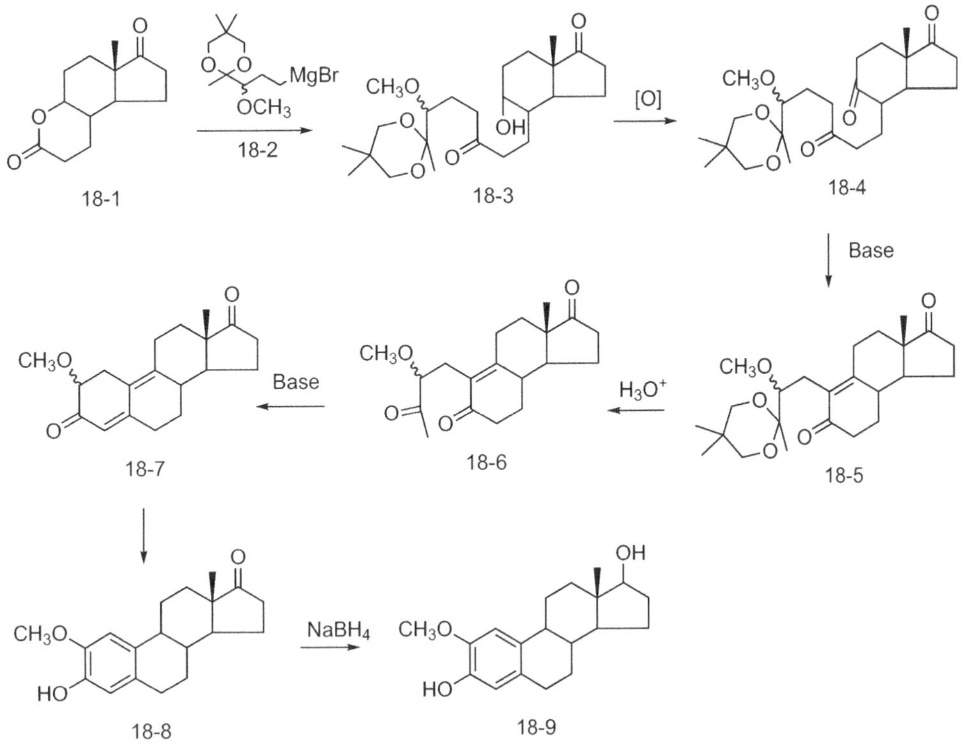

Scheme 3.18

3.3 Chemical Reactions of Estranes

3.3.1 Aromatic A-ring Reactions

3.3.1.1 Nitration

Reaction of a solution of estrone in acetic acid with nitric acid results in the formation of approximately equal amounts of 2- (**19-1**) and 4-nitroestrane (**19-2**). Each of those products can then be converted to the corresponding nitroestradiol, **19-3** and **19-4**, by means of sodium borohydride (Scheme 3.19). Direct nitration of estradiol under the same conditions leads to roughly the same ratio of 2- and 4-nitroestradiol.

Different results are obtained when this reaction is carried out using metal nitrates on solid supports. Reaction of a solution of estrone in THF with montmorillonite-supported bismuth nitrate gives a 2:1 ratio of 4-nitroestrone (**19-2**) to 2,4-dinitroestrone (**20-1**) (Scheme 3.20).

The chemical properties of nitroestrones are much the same as those of simpler aromatic compounds. The nitro group in **19-1** can, for instance, be converted to the aniline **21-1** for example by reduction with tin and acid; treatment of the resulting aniline with nitrous acid then gives the diazonium salt **21-2** (Scheme 3.21). Photolysis of that salt in methanol leads to formation of 2-methoxyestrone, (**21-3**), in this case confirming the structure of one of the metabolites of estradiol.

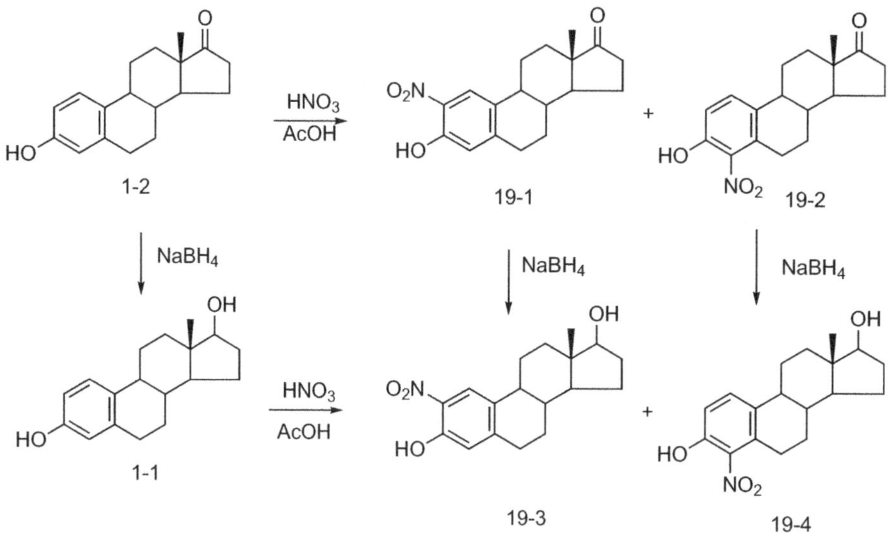

Scheme 3.19

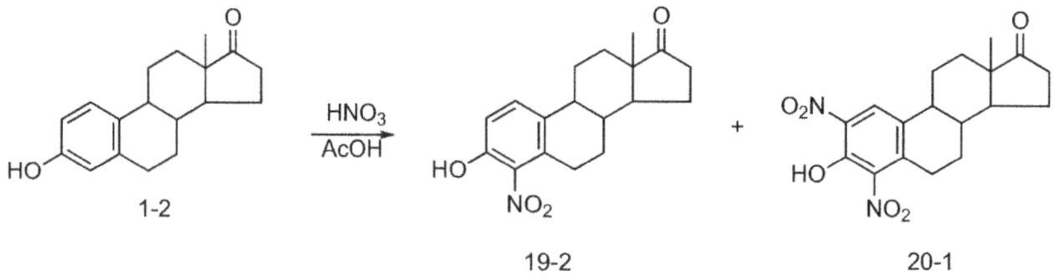

Scheme 3.20

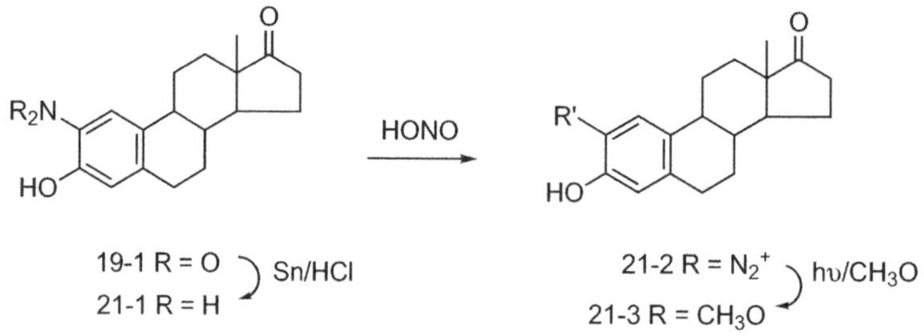

Scheme 3.21

3.3.1.2 Oxidation Reactions

The phenolic ring in estrone can be readily oxidized. Reaction of estrone with Fremy salts (peroxylamine disulfonate) affords a mixture of the two isomeric catechols. In a more controlled manner, treatment of estrone with 2-iodoxybenzoic acid (**22-1**) leads intially to a mixture of the 2,3-quinone (**22-2**) and its 3,4-isomer (**22-3**) (Scheme 3.22). These products are then reduced *in situ* with ascorbic acid to afford 2-hydroxyestrone (**22-4**) and 4-hydroxyestrone (**22-5**).

3.3.1.3 Carbon–Carbon Bond Formation

Alkylation of the hydroxyl group in estrone with allyl bromide leads to the allyl ether **23-1** (Scheme 3.23). Heating a solution of that ether in refluxing *N,N*-diethylaniline leads to the corresponding products from Claisen rearrangement. The product from this reaction, as in the case of nitration and oxidation, also leads to a mixture of regioisomers. Although one might expect the less hindered isomer **23-2** to predominate, the 4-alllylestrone (**23-3**) is actually formed in a ratio of almost 2:1 over 2-allylestrone (**23-2**).

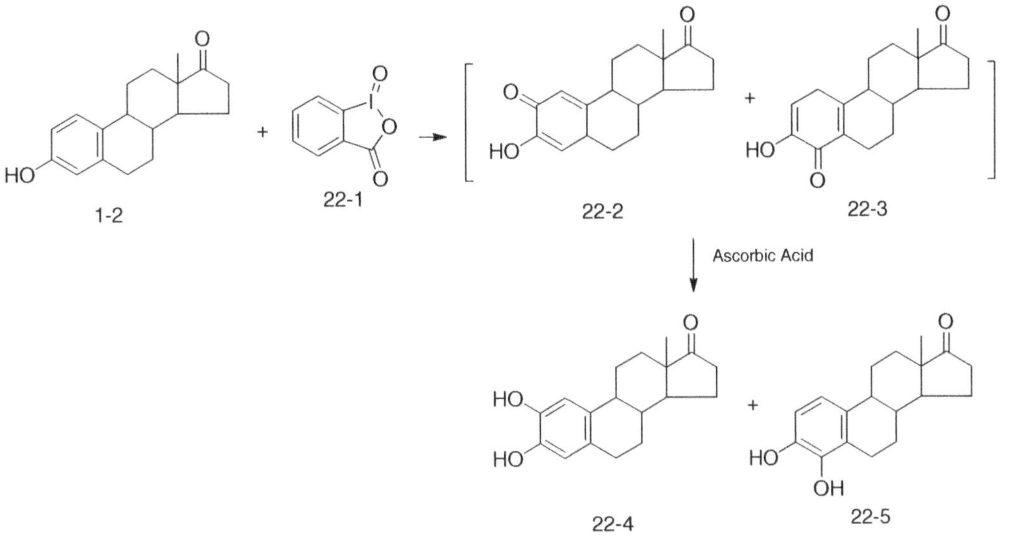

Estranes: Steroids in Which Ring A is Aromatic **39**

Scheme 3.22

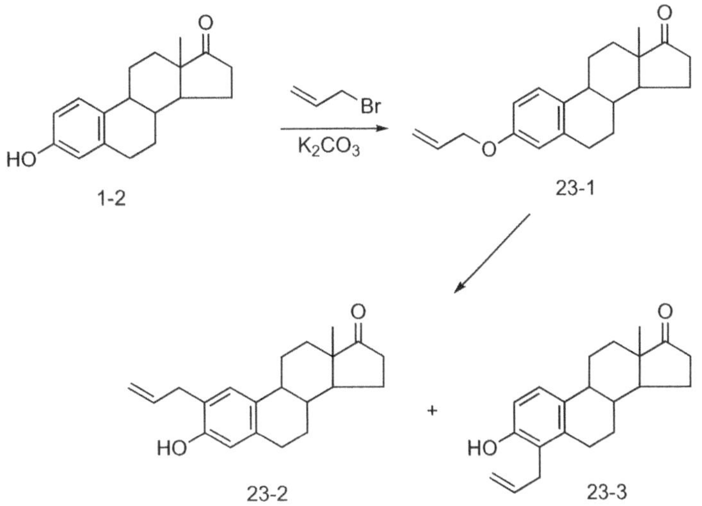

Scheme 3.23

Palladium-catalyzed carbonylation reactions provide a means for replacing a phenolic carbon–oxygen bond by a carbon–carbon bond. In the case at hand, the hydroxyl group in estrone is first converted to its trifluoromethylsulfonyl (triflate) derivative by reacting the compound with triflic anhydride in the presence of a hindered base (**24-3**) (Scheme 3.24). That derivative is then condensed with carbonyl:diisopropylamine in the presence of palladium

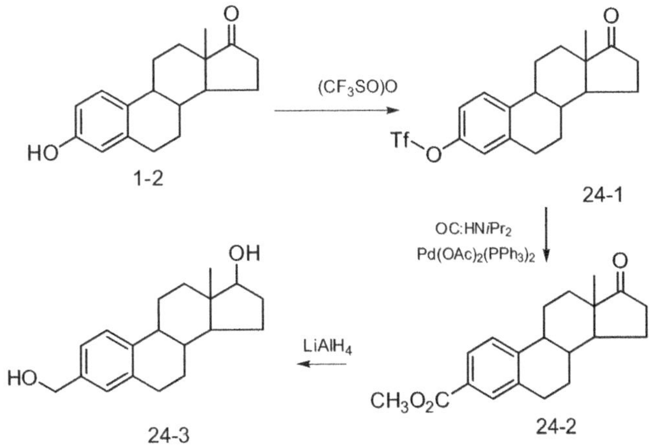

Scheme 3.24

diacetate bistriphenylphosphine. This affords the methyl ester of a derivative of estrone that now carries a carboxylic acid at C_3 (**24-3**). Reaction of that product with lithium aluminum hydride reduces both the ester group and the ketone at position 17 to yield the estradiol homologue **24-4**.

3.3.1.4 Deoxygenation

It is possible, finally, to remove oxygen at position 3 altogether. This transformation starts by conversion of the hydroxyl at position 3 in estrone to its phosphite derivative, **25-1**, by reaction with diethylphosphite chloride (Scheme 3.25). The resulting derivative is the treated with lithium in liquid ammonia. This reaction affords the deoxy derivative **25-3**; the ketone at C_{17} is reduced to a β-hydroxyl group under the reaction conditions.

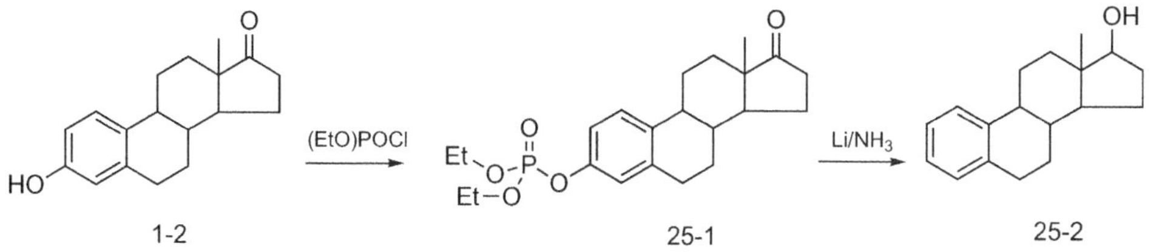

Scheme 3.25

3.3.2 Modifications on Ring B

Reaction of estrone methyl ether with 2,2-dimethylpropane-1,3-diol in the presence of a catalytic amount of acid leads to derivative **26-1**, in which the ketone at 17 is protected as an acetal (Scheme 3.26). Treatment of this intermediate with pyridinium chlorochromate leads to oxidation of the C_6 benzylic carbon atom to a carbonyl group (**26-2**). Potassium *tert*-butoxide abstracts a proton from the adjacent methylene at C_7; alkylation of the resulting anion with 4-(*N,N*-dimethyl)butyl iodide gives **26-3** as a mixture of diastereomers. The carbonyl group is next reduced to an alcohol by means of sodium borohydride (**26-4**). Dehydration of the newly introduced hydroxyl group is arguably facilitated by the adjacent aromatic ring (**26-5**). Aqueous acid removes the 17-acetal to afford **26-6**, which is in essence an equilinin derivative.

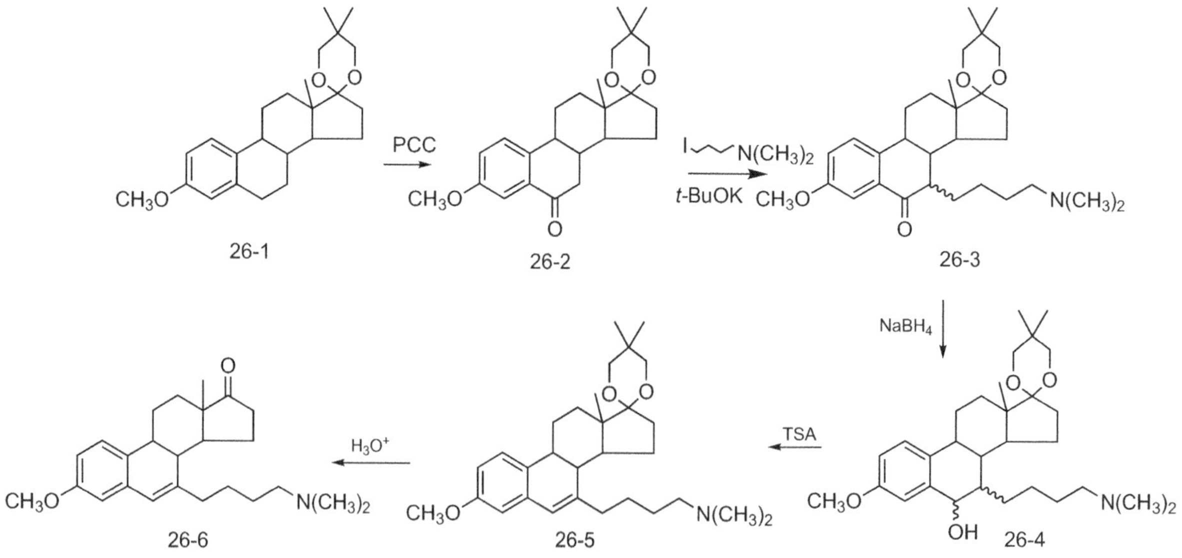

Scheme 3.26

Preparation of estranes bearing substituents at position 6 relies on the availability of Δ^6 starting materials. In a typical example, 17-acetal-protected 6-dehydroestrone (**27-1**) is treated with an oxidizing agent such as peracetic

Estranes: Steroids in Which Ring A is Aromatic 41

acid. to afford the epoxide **27-2** from attack on the more open backside (Scheme 3.27). Reaction of that intermediate with the Grignard reagent from methyl bromide opens the oxirane. The regiochemistry of this transformation follows diaxial opening of the oxirane to alcohol **27-3**. The newly introduced hydroxyl is then converted to its mesylate, **27-4**, with methanesulfonyl chloride. Collidine-mediated elimination of methanesulfonic acid then restores the unsaturation (**27-5**). Hydrolysis removes both the acetal and the acetate at position 3 to afford 6-methylequilinin (**27-6**).

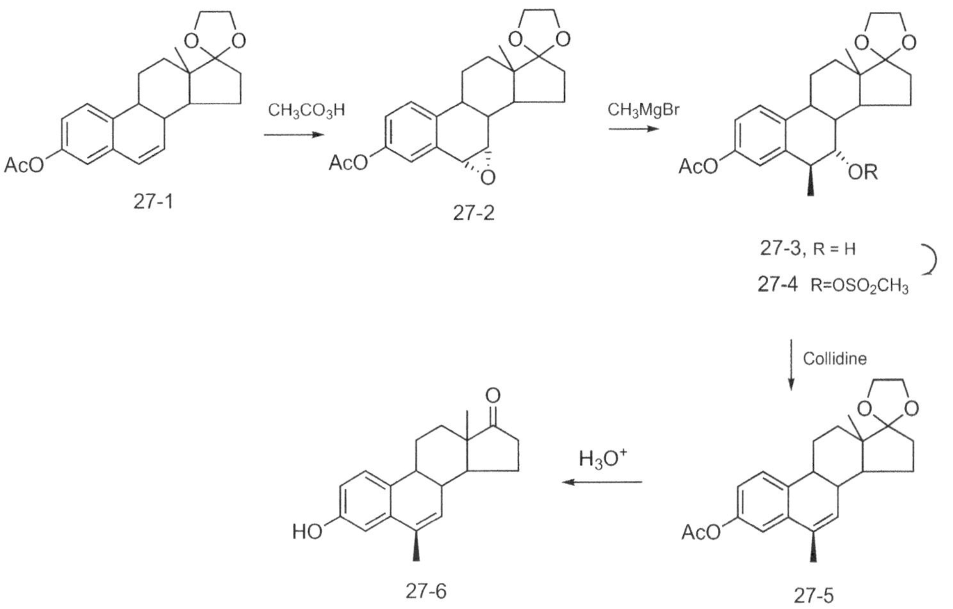

Scheme 3.27

3.3.3 Modifications on Ring C

The oxidizing agent chloranil (**28-1**) has been used for removing hydrogen from a range of hydroaromatic systems; the reagent is itself reduced in the process to a hydroquinone (**28-2**) (Scheme 3.28). The more electronegative analogue dichlorodicyanoquinone (DDQ) (**28-3**) has found extensive use for introducing further unsaturation into ring A in preganes (see Chapters 6 and 7).

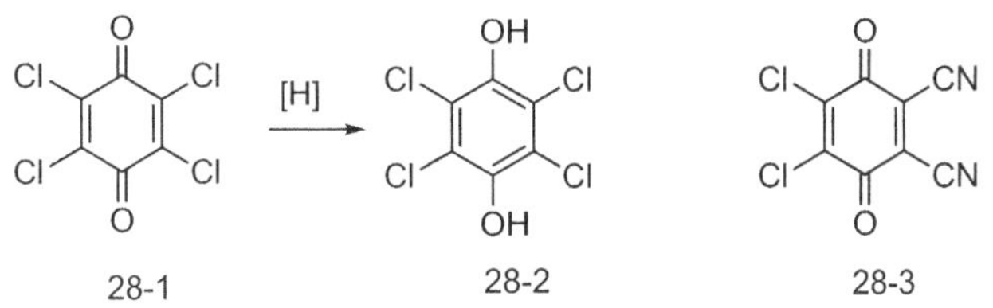

Scheme 3.28

Oxidation of the 17-acetal of estrone (**29-1**, from estrone and propylene-1,3-glycol) with DDQ abstracts hydrogen from ring C to form the styrenoid derivative **29-2** (Scheme 3.29). The driving force behind this particular regiochemistry may be related to the more transoid nature of that compound compared with the 6,7 isomer. Hydroboration of the product followed by oxidative workup affords the 11α-hydroxy derivative **29-3**. Oxidation of the newly introduced hydroxyl group with chromium trioxide followed by hydrolysis of the 17-acetal yields 11-oxaestrone (**29-5**).

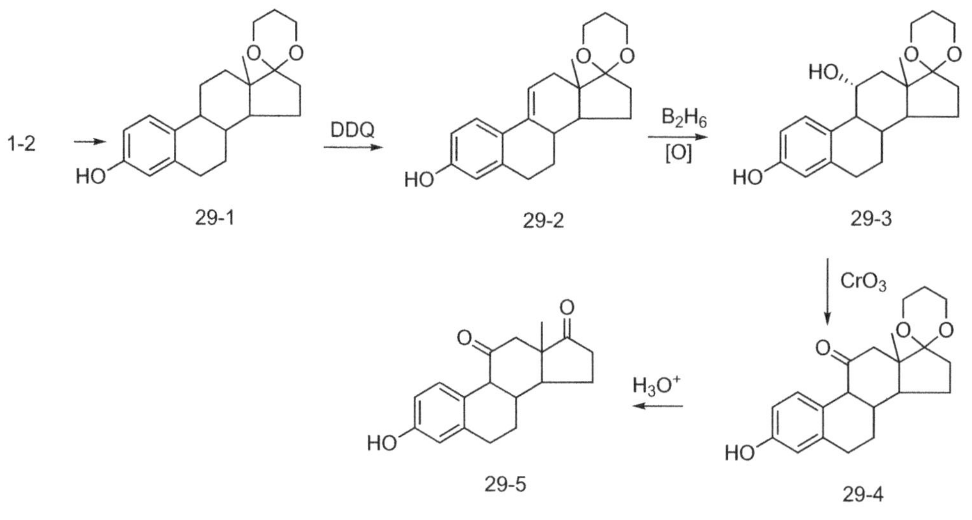

3.3.4 Modifications on Ring D

3.3.4.1 Ring Modification

Estradiol and estrone are metabolized to an array of oxidized products, one of which consists of the 16α-hydroxy derivative **30-4**. One approach to preparing that compound starts by reaction of estrone with isopropylidene acetate, to afford the acetate of the enolic form of the ketone and also the ester of the phenol at position 3 (**30-1**) (Scheme 3.30). Treatment of that product with perbenzoic acid leads to the α-oxirane **30-2**, formed from approach of the reagent from the less hindered backside. Acetolysis of that intermediate gives 16α-acetoxyestrone (**30-3**). Reaction of that product with lithium aluminum hydride leads to reduction of the 17-carbonyl and also the phenolic ester to give the *trans*-diol 16α-hydroxy-17β-estradiol (**30-4**). The same product is obtained on reducing **30-2** directly also with lithium aluminum hydride.

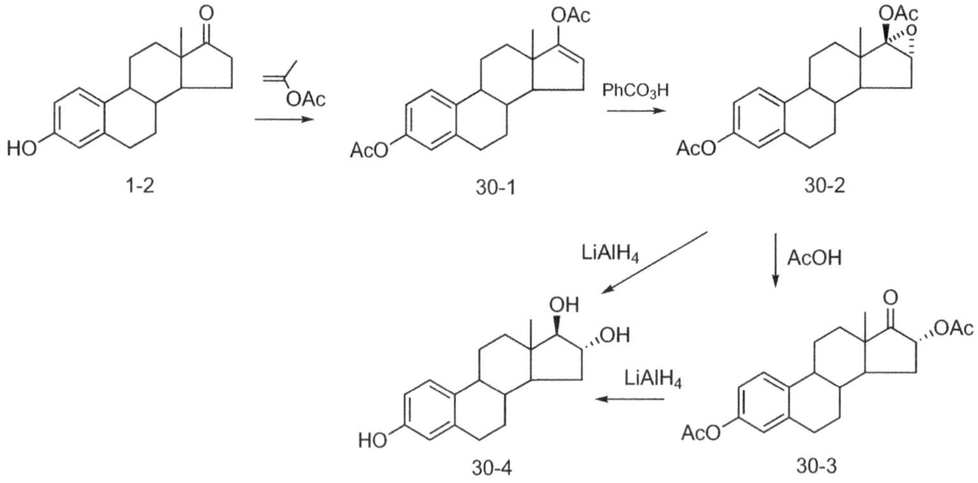

Preparation of the corresponding *cis*-16α,17α-diol relies on the known propensity of many inorganic oxidizing agents to produce *cis*-diols. The sequence for synthesizing that isomer starts with heat-induced elimination of the elements of benzoic acid from estradiol 17-benzoate. The resulting 16-dehydro compound, **31-2**, is then treated with osmium tetraoxide (Scheme 3.31). The stereochemistry of that transformation can be rationalized by positing the intermediacy of a complex such as that depicted in **31-3**. The overall result is the formation of the *cis*-diol 16α-hydroxy-α-estradiol (**31-4**).

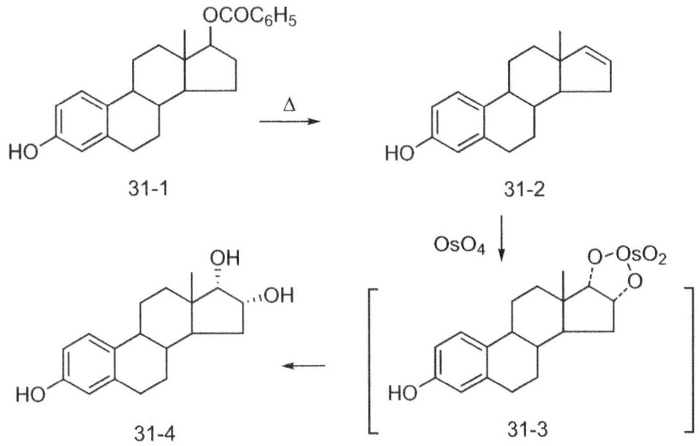

Estranes: Steroids in Which Ring A is Aromatic

Scheme 3.31

A slight modification of Corey synthesis (see Scheme 3.16) affords an estrane that bears a hydroxyl at C_{14}. The 14-dehydroestrone intermediate from that synthesis is first reduced to give the corresponding 17β-hydroxy analogue. This, in turn, is converted to its tert-butyldimethylsilyl ether (TBDMS) (**32-1**) by reaction with the silyl chloride (Scheme 3.32). Oxidation by means of *m*-chloroperbenzoic acid (mCPBA) affords the 14–15 epoxide **32-2** as a 3:1 mixture of α- and β-epimers. Treatment of the former with lithium aluminum hydride leads to the alcohol **32-3**. Mild acid then cleaves off the silyl protecting group. The alcohol at position 17 is then oxidized with Jones' reagent (chromium trioxide in acetone) to afford 14α-hydroxyestrone 3-methyl ether (**32-5**).

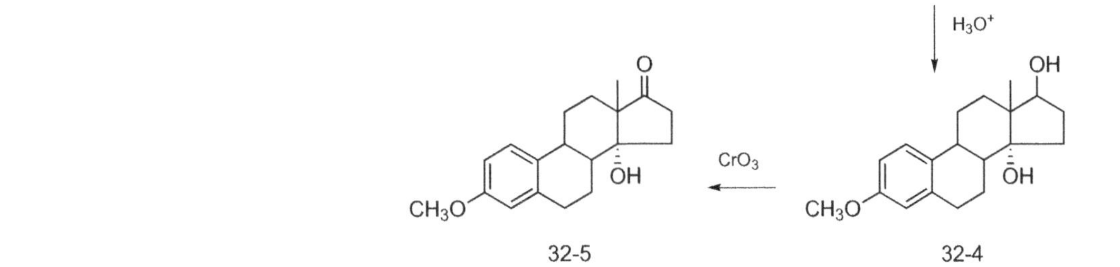

Scheme 3.32

3.3.4.2 Modifications at C_{17}

The ketone at position 17 readily adds a wide selection of nucleophiles. The crowded milieu on the β-face of ring D generally leads to the addition of nucleophiles from the more open α-face of the ketone.

Reaction of estrone methyl ether **13-2** with methyl Grignard reagent affords the corresponding 17-alkylated alcohol **33-2** (Scheme 3.33). Addition of acetylene is notable for the mild reaction conditions. A good yield of **33-1** is obtained on simply bubbling acetylene into a solution of the steroid and potassium hydroxide. This product, dubbed mestranol, is a potent estrogen present in very small quantities in the great majority of oral contraceptive products.

A variation of the Suzuki-type coupling reaction provides another means for placing a functionalized carbon atom on to position 17. The sequence starts with conversion of estrone methyl ether to its hydrazide by reaction with hydrazine. Treatment of this intermediate with 1,1,3,3 tetramethylguanidine (TMG) and iodine leads to the 17-enol

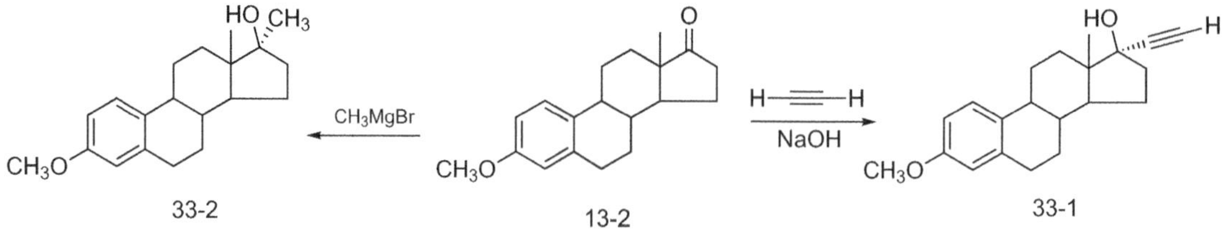

Scheme 3.33

iodide **34-2** (Scheme 3.34). The crucial step in the sequence comprises reaction of a solution of the iodide in methanol with carbon monoxide in the presence of palladium acetate–triphenylphosphine complex. This transformation affords the 17-carbomethoxy derivative, **34-3**, of estrone. Replacing methanol with amines leads to the corresponding carboxamides.

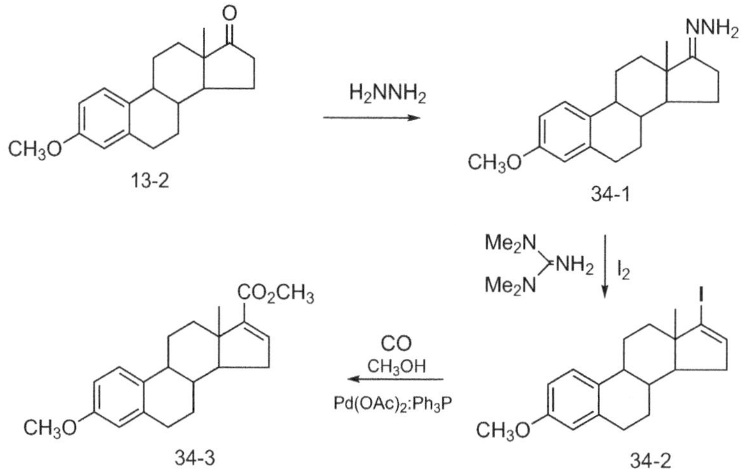

Scheme 3.34

Estrone methyl ether (**13-2**) readily forms an oxime (**35-2**) on reaction with hydroxylamine (Scheme 3.35). That intermediate can then be taken on to 17β-aminoestrone 3-methyl ether (**35-3**) by metal hydrides such as lithium aluminum hydride. In a variation on that theme, reaction of the same starting material (**13-2**) with methylamine affords

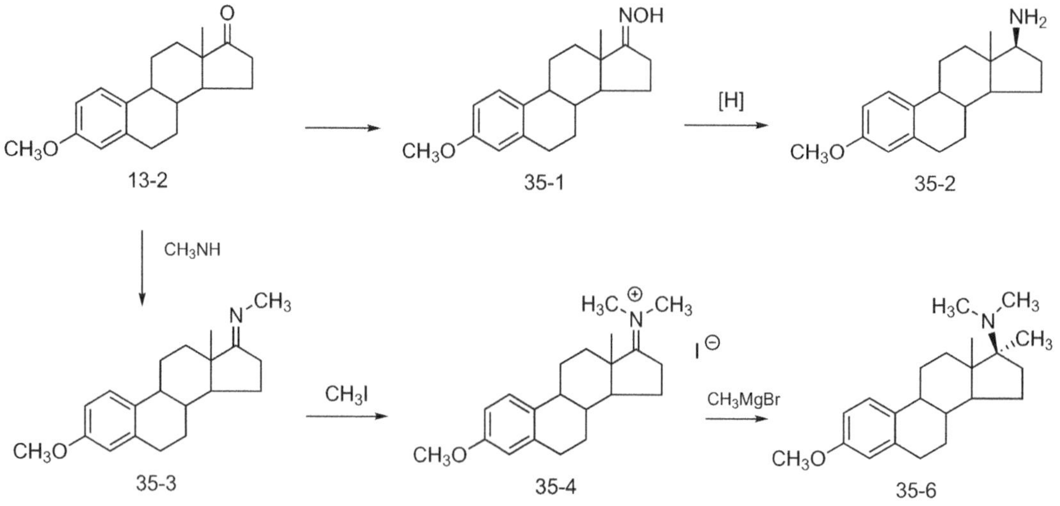

Scheme 3.35

the methylimino derivative **35-3**. This derivative is next activated towards nucleophiles by conversion to the iminium salt (**35-4**) by alkylation with methyl iodide. Reaction of that salt with methylmagnesium bromide leads to addition at position 17 (**35-6**). Assignment of the stereochemistry of the newly introduced methyl group as α rests on analogy with the outcome of addition to 17-ketones.

3.4 Some Drugs Based on Estranes

As noted at the outset of this chapter, frank estrogens such as estradiol (**1-1**), and also the mixture of sulfated estrogens from pregnant mare's urine, are often prescribed for easing unwanted side effects in women at menopause. Estradiol itself is poorly absorbed from oral dosage and is in addition readily inactivated in the liver. One approach to overcoming this drawback involves preparing esters of estradiol with very lipophilic fatty acids. Those esters are then dissolved in vegetable oils and the solutions injected just below the skin to form long-term depots. As these esters slowly leach into the circulation, esterase enzymes cleave the carbon–oxygen bond, releasing estradiol. Forcing conditions are required for esterifying estradiol due to steric hindrance at position 17. Mild hydrolysis of the first-formed 3,17-diesters obtained under those conditions affords the desired monoesters (Scheme 3.36). Typical examples include estradiol benzoate (**36-2**) and estradiol cypionate (**36-3**).

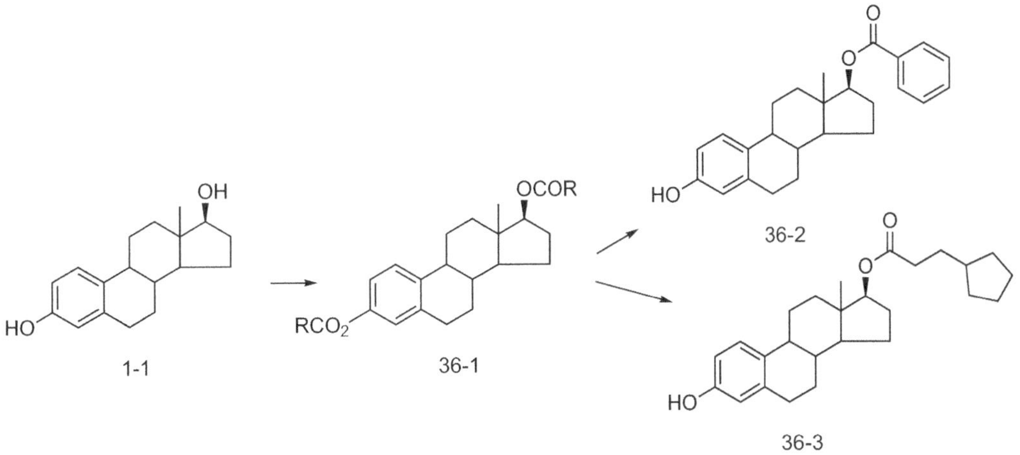

Scheme 3.36

The secondary carbon at C_{17} is one of the principal sites for metabolic inactivation of estrogens. Adding another substituent at that position extends the biological half-life of orally active drugs. Reaction of estrone (**1-2**) with lithium acetylide thus leads to the potent estrogen ethynylestradiol (**37-1**) (Scheme 3.37). In similar vein, the reaction of 2-iodofuran with an akyllithium gives the lithio reagent of furan by halogen interchange. The resulting reagent adds to the ketone in **1-2** to yield estrofurate **37-2**. Both reactions are highly stereospecific in that the product consists virtually entirely of the α-adduct, pointing once more to the hindrance about the β-face of the steroid.

A significant proportion of breast cancers contain estrogen receptors. Receptor binding of estrogen at those sites stimulates the growth of the tumor. Various chemotherapeutic approaches attempt to steer alkylating drugs selectively to those receptors in order to diminish the harm from indiscriminate addition of alkyl groups to normal tissues. Attaching an alkylating function to estrone was designed to deliver that group selectively to estrogen receptors.

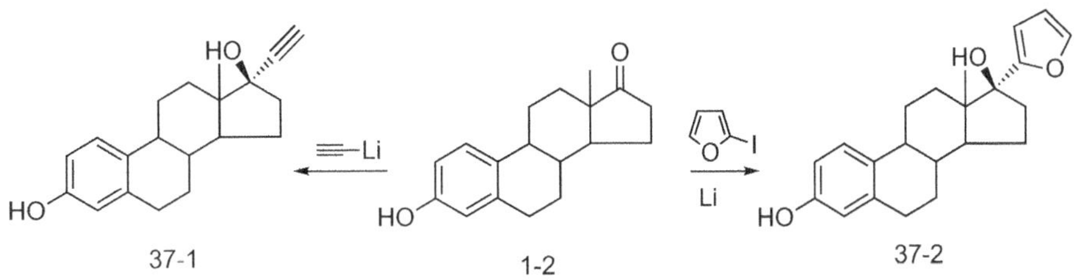

Scheme 3.37

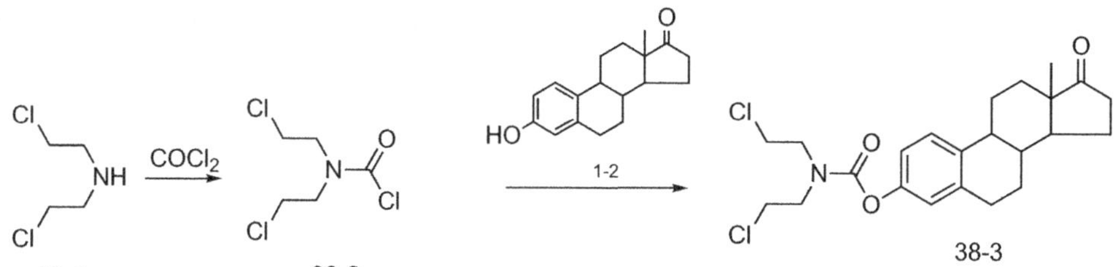

38-1 38-2 38-3

Scheme 3.38

The first step in this concise synthesis comprises reaction of the alkylation moiety **38-1**, also known as 'nitrogen mustard', with an excess of phosgene to afford the carbamoyl chloride **38-2** (Scheme 3.38). Reaction of that intermediate with estrone in the presence of base yields the anticancer drug estramustine (**38-3**).

4
Gonanes or 19-nor-Steroids

The gonanes, like the structurally closely related estranes, have no counterpart in the plant kingdom. All compounds lacking the angular carbon at C_{10} are consequently prepared from other steroids or alternatively by total synthesis. Gonanes were thus originally prepared from estranes by means of one of several versions of the Birch reduction. This reaction provides a convenient means for converting aromatic rings such as ring A in estranes to di- or in some cases tetrahydrocyclohexanones. The total synthesis approach to gonanes, developed later, affords the means for preparing a wider range of modified gonanes. The biological activity of compounds in this class falls into two categories. Those gonanes that exhibit minor modifications at or near C_{17} tend to show androgenic–anabolic activity, whereas the gonanes that bear substituents that comprise chains of two or more carbon atom chains on C_{17} usually act as progestins. The decidedly unnatural compounds that bear a modified benzene ring at C_{11} act antagonists at glucocorticoid and more significantly progesterone receptor sites.

4.1 Preparation of Gonane Starting Materials
4.1.1 Birch Reduction

The Birch reduction comprises a means for adding two hydrogen atoms to an aromatic ring by means of a metal, most often lithium, and an alcohol in liquid ammonia as solvent. A co-solvent, often tetrahydrofuran (THF), is often added due to the very poor solubility of steroids in ammonia. The use of the more expensive sodium was at one time precluded because traces of iron in that metal catalyzed the conversion of the metal to the strong base sodium amide. Very pure sodium, free of iron impurities, is now used for commercial-scale reductions.

The first step in this transform comprises formation of a radical dianion (**1-2**) from reaction of at the sites of lowest electron density (Scheme 4.1). In the case of steroids, this requirement rules out all but positions 1,4 of the aromatic ring A. The charges on the intermediate then strip protons from the alcohol in the reaction medium to yield the dihydrobenzene **1-3**. The alcohol, frequently *tert*-butanol, thus acts as a quench. The enol ether function in the product is sufficiently robust to serve as a protecting group in many subsequent reactions. Mild acid hydrolysis of that enol ether affords the surprisingly stable unconjugated ketone **1-4**. Treatment of this last intermediate with mineral acid causes the unsaturation to move over into conjugation at C_4 (**1-5**).

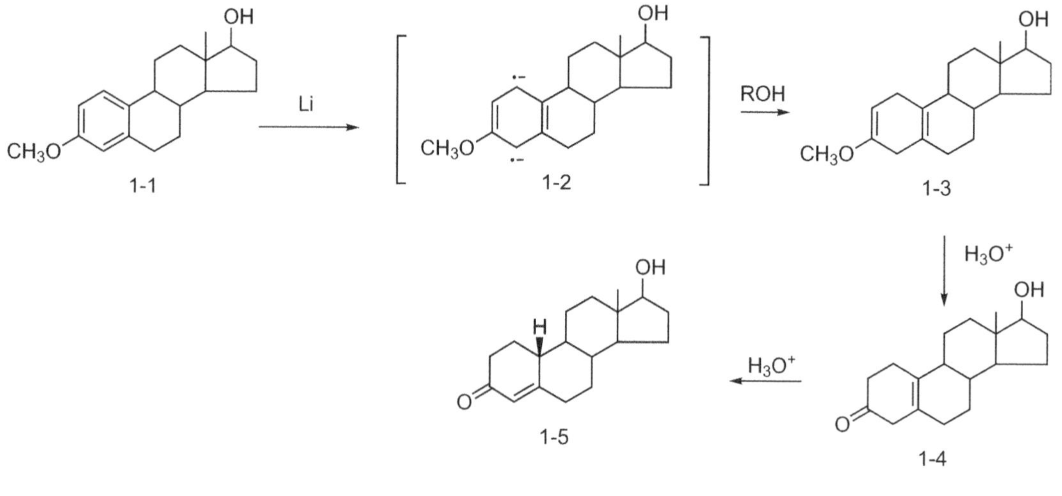

Scheme 4.1

4.1.1.1 13-Ethylgonanes

The methyl group on position 13 is replaced by an ethyl fragment in a significant number of gonanes used as drugs. The key intermediate (**2-3**) in the synthesis of those compounds is obtained by replacing the methyl cyclopentadione by its ethyl analogue, **2-2**, in the Smith–Torgov procedure (Scheme 4.2). That steroid is then taken on to the gonane **2-4** in several steps, the last of which comprises Birch reduction.

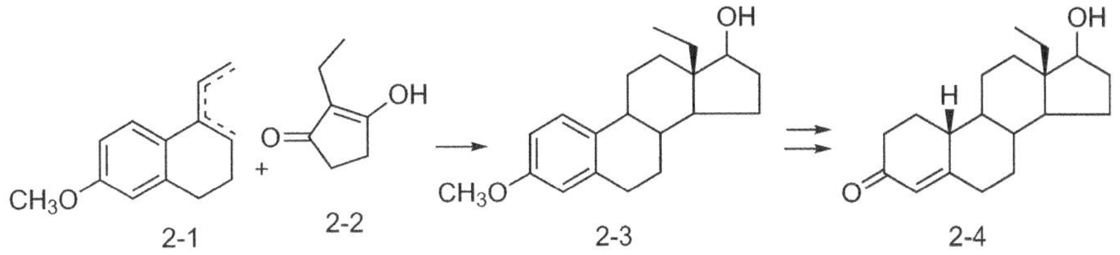

Scheme 4.2

4.1.2 Synthesis by Sequential Annulation Reactions

The Smith–Torgov synthesis is somewhat limited in providing steps or sites for introducing structural variation. A lengthier, although more flexible, procedure actually starts with an open-chain counterpart of the tetralone used in the Smith–Torgov synthesis. The presence in the scheme of an intermediate carboxylic acid (**3-3**) in addition allows for efficient optical resolution at an early stage. This opportunity for early resolution avoids the prospect of having to discard half of the compound that has gone through many steps.

The sequence starts with the conjugate addition of 2-methylcyclopenta-1,3-dione (**3-2**) to the vinyl ketone **3-1** (Scheme 4.3). The adduct then cyclizes to the hydrindone **3-3**. Catalytic hydrogenation of the double bond in what will become ring C proceeds by addition of hydrogen from the side away from the angular methyl group. The product **3-4** now incorporates the *trans* C–D ring fusion required for biological activity. The next several steps (not depicted) involve selective protection of the more electrophilic cyclohexanone carbonyl group as an acetal, hydride reduction of the cyclopentyl carbonyl group, removal of the acetal and finally saponification to yield acid **3-5**. Resolution can be carried out at this stage using, for example, optically active bases. Reaction with benzoyl chloride then affords the corresponding benzoate, **3-6**. Acetic anhydride then causes the keto acid to dehydrate to form unsaturated enol lactone **3-7**. Reaction of that lactol with Grignard reagent **3-8** converts the enol lactone to a cyclohexenone. The acetal group in the side chain is then hydrolyzed to the free ketone **3-9**.

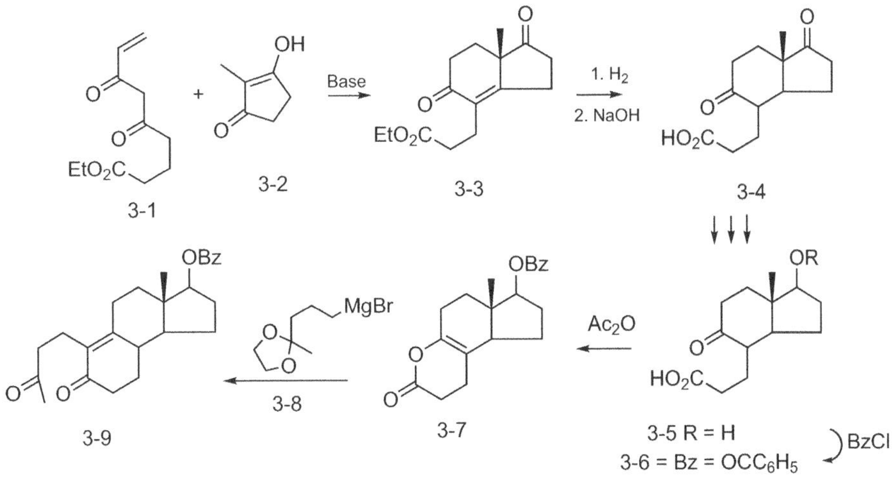

Scheme 4.3

This complex transform can be envisaged by assuming that one equivalent of **3-8** attacks the enol lactone carbonyl group to afford the hydroxylactone **4-2** (Scheme 4.4). This unstable intermediate actually comprises a hemiacetal of an enol lactone. The reaction medium is sufficiently basic to cause the underlying diketone (**4-3**) to undergo aldol cyclization to form cyclohexenone **4-4**.

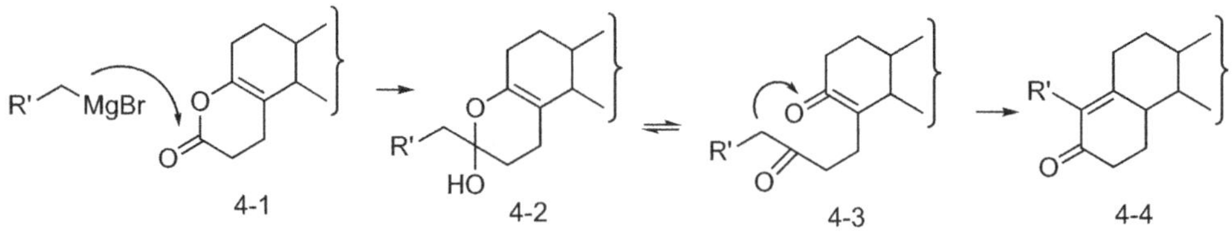

50 *Steroid Chemistry at a Glance*

Scheme 4.4

The methyl ketone at the end of the chain (**3-9**) can be expected to be more reactive than the ring enone. Reaction of **3-9** with piperidine thus forms an enamine from that carbonyl group **5-1**; this intermediate then cyclizes to the new enamine **5-2** (Scheme 4.5). This derivative now displays a gonane carbon skeleton. Hydrolysis of that compound with weak acid causes the enamine to hydrolyze, yielding the gonane **5-3**. This compound provides an additional site for attaching novel functions such as the double bond in ring C. Hydrolysis of **5-3** with strong acid goes on to **5-4**, in which the ketone is conjugated.

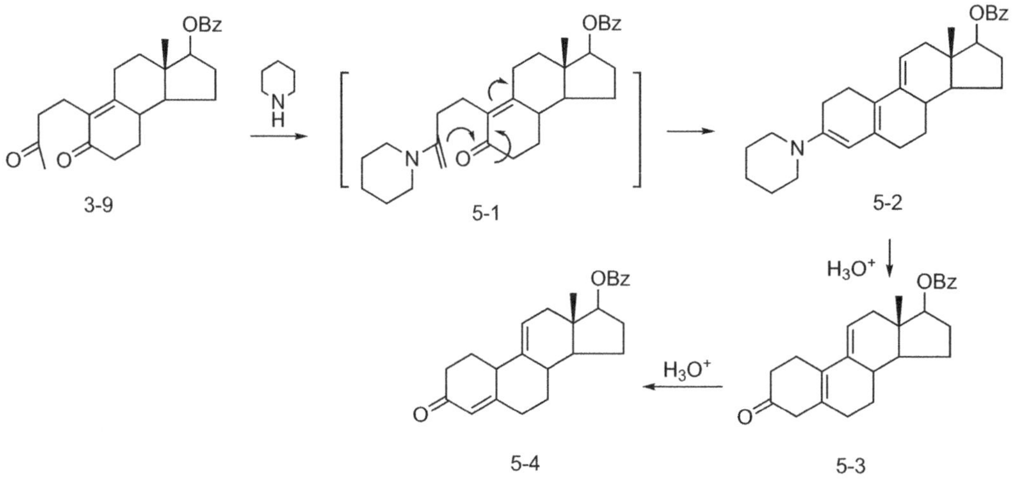

Scheme 4.5

4.2 Anabolic–Androgenic Gonanes

4.2.1 Biological Activity

Widely touted testosterone, in fact, serves mainly as an endogenous starting material for 5α-dihydrotestosterone, the substance actually responsible for biological activity by binding to the androgen receptor (Figure 4.1). 5α-Dihydrotestosterone holds a place in the male of the species roughly equivalent to that of estrogen in females. That compound and its synthetic counterparts support male reproductive function. Testosterone, which is secreted by the testes, maintains male features such as libido and production of sperm. That steroid and many of its analogues also stimulate growth of muscle tissue. This anabolic activity makes these compounds attractive to athletes seeking to improve their performance. The List of Controlled Substances published by the US Drug Enforcement Administration (DEA) contains no fewer than 55 proscribed anabolic–androgenic compounds. It was believed at one time that 19-nor analogues of testosterone showed increased anabolic over androgenic activity. It now seems that the gonanes in this

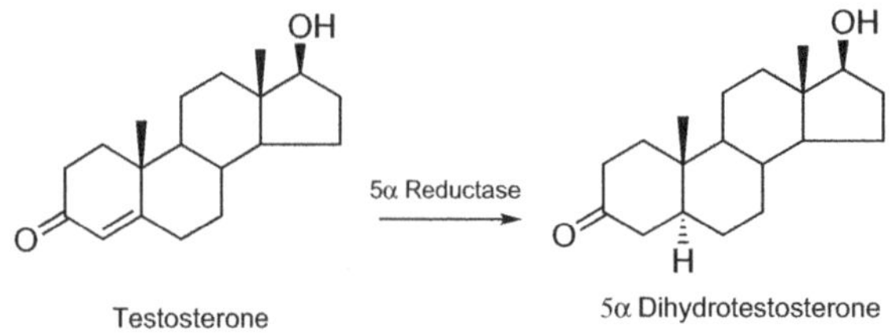

Figure 4.1 Testosterone and 5α-dihydrotestosterone

chapter have much the same ratio of anabolic to androgenic activity as testosterone itself. The attraction of these compounds to the gray market lies in the hope that they may avoid detection by gas–liquid or high-performance liquid chromatography.

4.2.2 Synthesis of 19-Norandrogens

The product from Birch reduction of estradiol 3-methyl ether, nandrolone, **1-5**, comprises one of the simplest androgens in the gonane series. This compound suffers extremely fast metabolic inactivation by attack at C_{17}; most androgens consequently incorporate an additional substituent at that position.

4.2.2.1 Gonanes with Additional Carbon Atoms

The drug that bears a methyl group in addition to the 17β hydroxyl at C_{17} is prepared from estrone 3-methyl ether (**6-1**) (Scheme 4.6). The aromatic ring in this case serves as a latent cyclohexenone. Condensation of estrone 3-methyl ether with the Grignard reagent ethylmagnesium bromide gives the adduct **6-2**, which then is subjected to Birch reduction. The resulting 3-enol ether is then treated with acid. Cleavage of the ether with concomitant shift of the double bond into conjugation gives the androgen norethandrone (**6-3**).

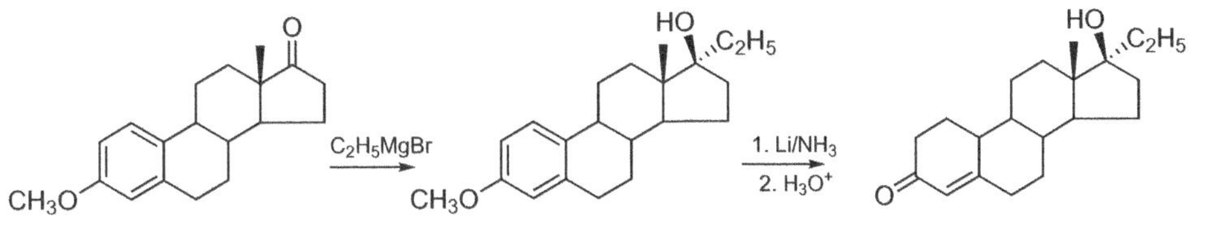

Scheme 4.6

The intermediate from Birch reduction of 13-ethylestrane (**2-3**, Scheme 4.2) serves as the starting material for another 19-norandrogen. Thus, Oppenauer oxidation of **7-1** leads to the 17-ketone **7-2** (Scheme 4.7). Reaction of the carbonyl at position 17 this time with ethylmagnesium bromide gives the product from addition of an ethyl group (**7-3**). Acid hydrolysis results in scission of the enol ether on ring A followed by shift of the olefin into conjugation. The product **7-4** is the androgen–anabolic agent norbolethone.

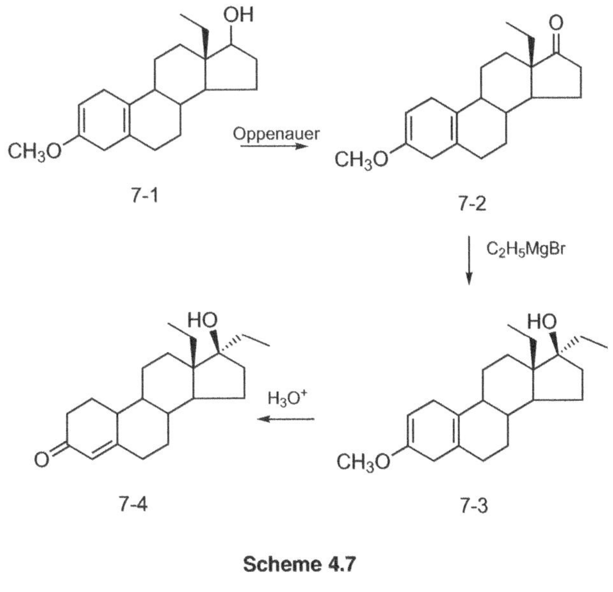

Scheme 4.7

Addition of an ethyl group at C_{16} is a much more complex proposition. The scheme for preparing the 16β-ethyl derivative begins with Knoevenagel condensation of DHEA acetate (**8-1**) with acetaldehyde and base to afford the ethylidene derivative **8-2** (Scheme 4.8a). Catalytic reduction adds hydrogen from the bottom side to give **8-3**. The carbonyl at C_{17} is then reduced with lithium aluminum hydride; the acetate is also reduced in the process. Reaction with acetic anhydride restores diacetate **8-4**. Treatment with the equivalent of hypochlorous acid adds a hydroxyl group at C_5 and chlorine at C_6; initial formation of a cyclochloronium ion regulates regiochemistry. Oxidation by means of lead tetraacetate then leads to attack on the methyl group across the ring by oxygen at C_6 to form an internal tetrahydrofuran ring (**8-6**).

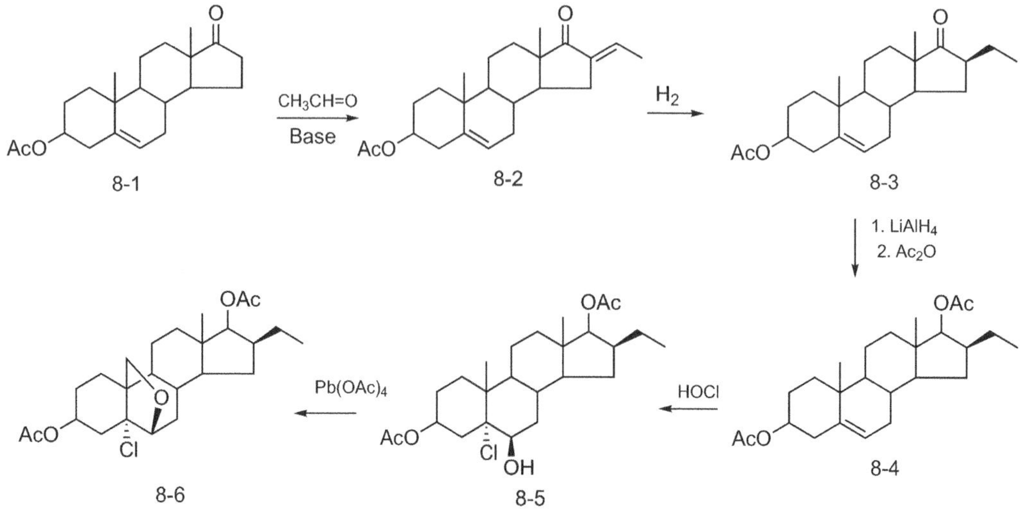

Scheme 4.8a

Selective saponification of the more accessible acetate at C_3 gives alcohol **8-7**, and oxidation of the newly revealed hydroxyl group using Jones' reagent affords ketone **8-8** (Scheme 4.8b). Treatment of the product from oxidation eliminates hydrogen chloride to form enone **8-9**. The key reaction to this scheme involves reductive opening of the furan ring to the 19-hydroxymethyl group (**8-10**). A second oxidation, this time with pyridinium chlorochromate (PCC), leads to the 19-aldehyde **8-11**. Base then leads to the expulsion the angular carbonyl group from what is in fact a vinylogous β-dicarbonyl array. Saponification subsequently cleaves the acetate at C_{17}. The product, oxendolone (**8-12**), is interestingly described as an antiandrogen.

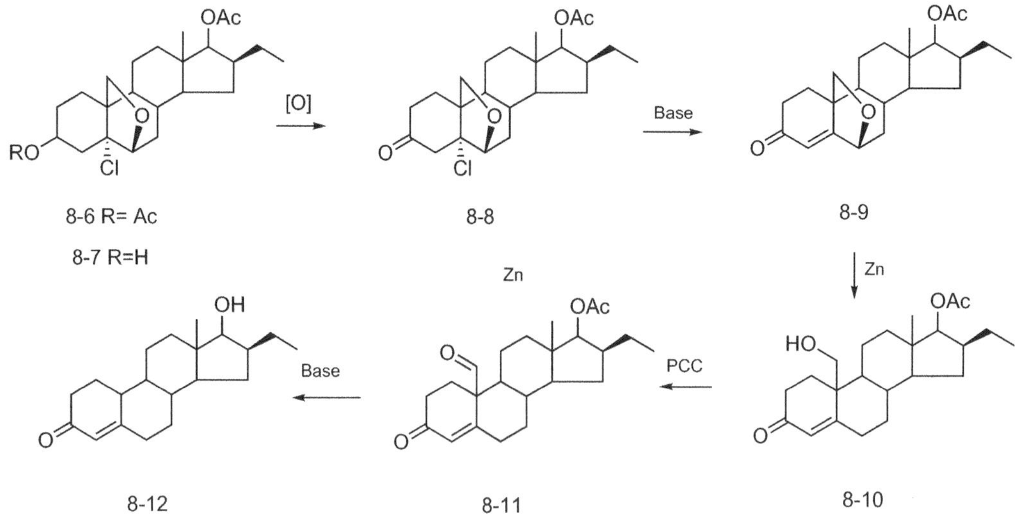

Scheme 4.8b

Adding a methyl group at C_7 is not quite as arduous a task as in the case of **8-12**. The first step involves extending the conjugation of the enone function by an additional double bond. Chloranil (tetrachloroquinone) is the forerunner of dichlorodicyanoquinone (DDQ), a reagent used extensively for introducing additional unsaturation in the progestin and corticoid series (see Chapters 6 and 7). In the case at hand, heating acetate **9-1** with chloranil gives the diene **9-2**, and reaction of that compound with methylmagnesium bromide in the presence of cuprous chloride leads to addition of the methyl group to position 7 at the end of the conjugated system (Scheme 4.9). The stereochemistry of the product (**9-3**) again illustrates the preference for additions from the backside. The alcohol at C_{17} is then oxidized to a ketone. Enamines are commonly used to activate adjacent functions; they are also not infrequently used, as in this case, as protecting groups. Thus, reaction of the intermediate **9-4** with pyrrolidine gives enamine **9-5**. This transformation emphasizes the clear difference in reactivity between ketones at C_7 and C_{17}. A second methyl Grignard addition gives the corresponding 17α-methyl derivative. Hydrolysis of the enamine function then affords mibolerone (**9-6**), a drug proscribed by the DEA.

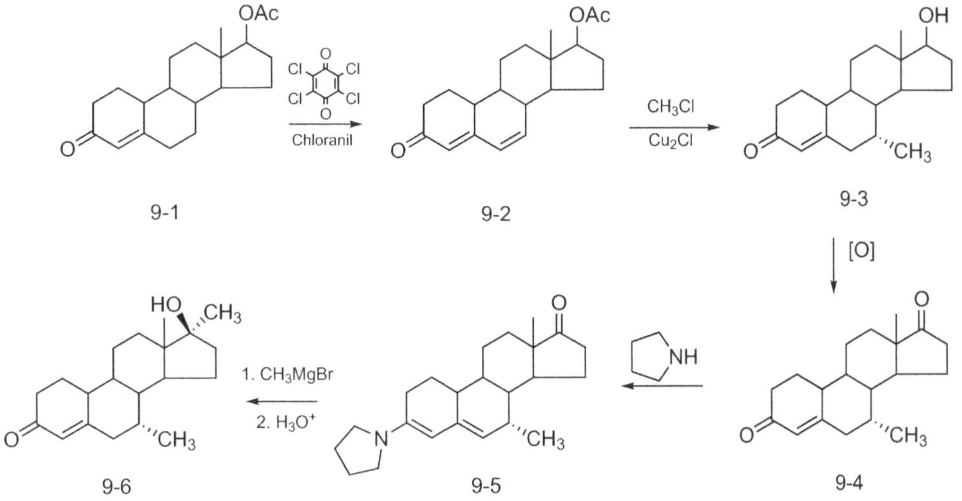

Scheme 4.9

Birch reduction of 1-methylestrone methyl ether (**10-1**), obtained by acid-catalyzed rearrangement of androsta-1,4-diene-4,17-dione, leads to the corresponding 1,4-dihdro intermediate **10-2** (Scheme 4.10). The carbonyl function at C_{17} is reduced to a hydroxyl group in the process. Warming a methanolic solution of the crude intermediate **10-2** in the presence of hydrogen chloride leads to scission of the methyl ether at C_3 with concomitant shift of the double bond to C_4. Scheme 4.10 depicts the methyl group at C_1 in the product in the more stable β configuration. Oxidation of the hydroxyl group on ring D with Jones' reagent then affords the dione **10-4**.

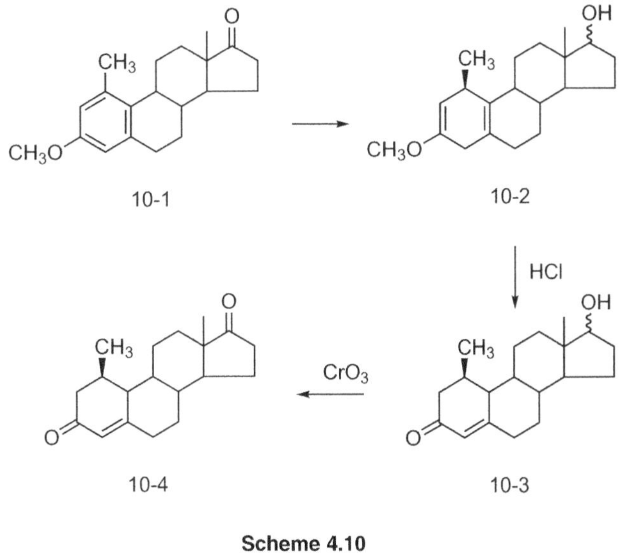

Scheme 4.10

The three additional carbon atoms at C_{10} in plomestane (**11-6**) technically remove that agent from the gonane series. The chemistry for its preparation is, however, more closely related to that of 19-nor compounds, hence its listing among gonanes. The synthesis that leads to that compound starts by careful oxidation of the hydroxyl group at C_{17} in the Birch reduction product **1-3** to afford the 3,17-diketone **11-1** (Scheme 4.11). Reaction of that intermediate with ethylene glycol catalyzed by a small amount of acid gives the bisacetal **11-2**. (In steroid chemistry, spiroacetals involving ethylene glycol or propylene-1,3-diol are more often designated as ketals.) That product is then treated with the equivalent of hypobromous acid (from *N*-bromosuccinimide in DMF), to give the 5,10-bromohydrin **11-3**. The regio- and stereochemistry can be rationalized by assuming that the first step involves formation of a 5,10-bromonium bridge between positions 5β and 10β; diaxial ring opening of that cyclic ion by hydroxide anion takes place by attack from the backside. Treatment with base then forms the epoxide **11-4** from displacement of bromide by the 5α-alkoxide. Reaction of that epoxide with the carbanion from trimethylsilylpropyne and butyllithium in the presence of cuprous anion opens the oxirane, this time from the β-face. The acetals are next removed by acid-catalyzed exchange with acetone. The acidic conditions also cause dehydration of the hydroxyl at C_5 and scission of the silyl protecting group all in one fell swoop. The product plomestane (**11-6**) inhibits the enzyme, aromatase, involved in the production of estradiol.

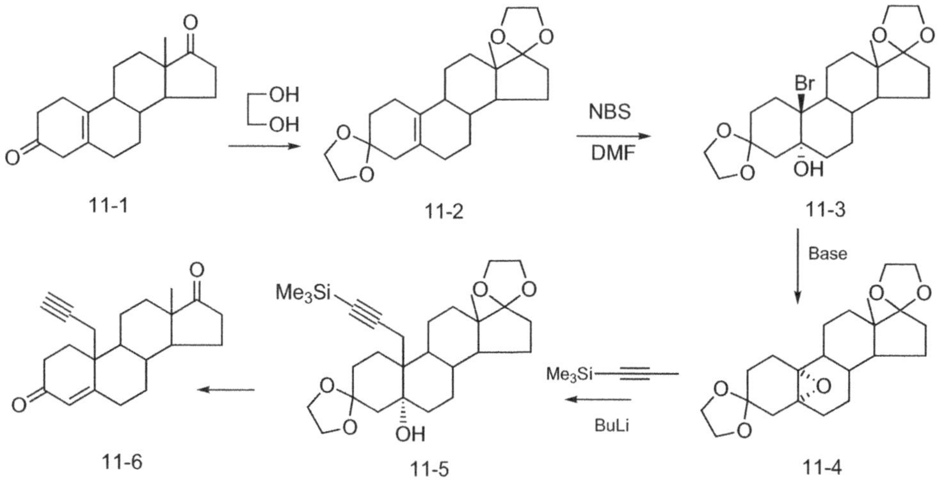

Scheme 4.11

4.2.2.2 Miscellaneous Gonanes

The greater potency of 2,3-dichloro-5,6-cyanoquinone as a dehydrogenation agent compared with that of chloranil has led to extensive use of this compound for introducing additional unsaturation in steroids. Treatment of the unconjugated diene **12-1**, from saponification of **5-3**, with DDQ gives the triene trenbolone **12-2** (Scheme 4.12), and in a similar vein, the corresponding gonane with a methyl group at C_{17} (**11-3**) gives the triene methyltrienolone (**12-4**), both potent anabolic–androgenic agents.

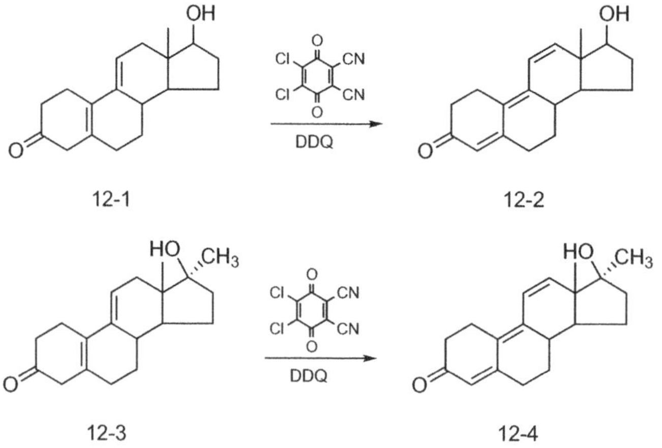

Scheme 4.12

Reaction of the unconjugated intermediate **13-1** with monoperphthalic acid (MPHA) yields a single 5,10-epoxide. The β stereochemistry of this product (**13-2**) follows from the same factors as those for the chloro- and bromohydrins described above. Reaction of the epoxide with hydrofluoric acid leads to diaxial opening of the oxirane and formation of the fluorohydrin **13-4** (Scheme 4.13). The course of the reaction of the epoxide with potassium hydroxide may be envisaged by assuming opening of the oxide to a 5,10-glycol. The transient hydroxyl at C_5 then dehydrates under the basic conditions to form an enone and thus **13-3**. Under the same conditions, the fluorohydrin **13-4** loses the elements of hydrogen fluoride to produce the same hydroxy ketone, **13-3**.

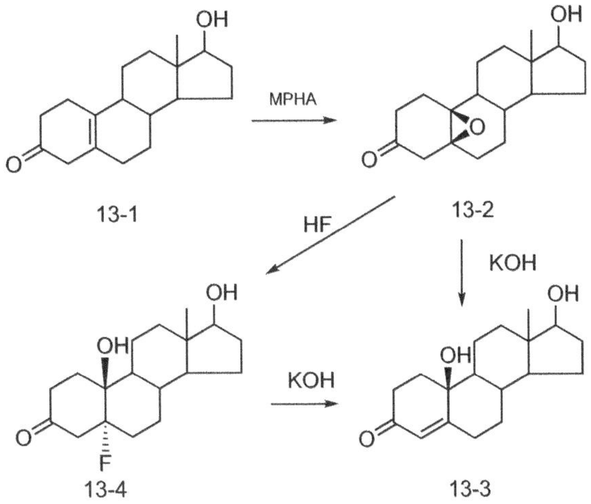

Scheme 4.13

4.3 Progestational Gonanes

4.3.1 Biological Activity

Progestins comprise the other hormone involved in the reproductive cycle in females. The concentration of progesterone (Figure 4.2) in blood, in common with that of the other reproduction-related hormone estradiol, shows a regular cyclic increase and the decrease each four-week period. Under control of the peptide hormone follicle-stimulating hormone (FSH), the cycle starts with the release of an egg from a blister-like structure on the ovary called a follicle. After the ovum has been released, the follicle ripens into the corpus luteum that then secretes progesterone. The increasing concentration of the hormone then prepares the inner wall of the uterus for implantation of a fertilized ovum should one appear. The presence of a fertilized ovum, implanted in the wall of the uterus, will cause the corpus luteum to persist until the end of pregnancy. This subsequent elevated progesterone levels halt further ovulation. In the absence of a fertilized ovum in the uterus, the corpus luteum gradually fades away and the lining of the uterus sloughs off, a process mirrored by a decline in progesterone levels. The ovulation-inhibiting activity of progesterone prompted research aimed at finding orally active progestins that would then act as oral contraceptives. Virtually every one of the currently marketed drugs in this class contains a progestin. The latter more often than not consists of an orally active 19-nor compound.

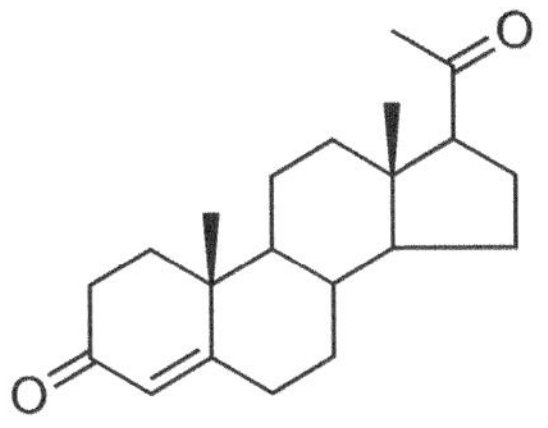

Figure 4.2 Progesterone

4.3.2 Preparation of 19-Norprogestins

4.3.2.1 Modification at C_{17}

The structural difference between the preceding androgenic gonanes and the counterpart progestins is very slight. The former generally bear only hydrogen or at most an α-methyl group at C_{17} whereas progestins carry a fragment that

56 *Steroid Chemistry at a Glance*

comprises two or more carbon atoms. Many of the schemes for synthesizing drugs in this class implicitly start from estranes, treating the aromatic ring as a latent cyclohexenone.

The sequence for preparing the first oral contraceptives thus starts by Birch reduction of estradiol 3-methyl ether (**14-1**) to afford the 1,4-dihydro derivative **14-2** (Scheme 4.14). The enol ether in this product is sensitive to acid, mandating that the succeeding reactions be carried out under neutral or slightly basic conditions. The hydroxyl at C_{17} in **14-2** is thus oxidized by means of the Oppenauer reaction (cyclohexanone and aluminum isopropoxide) to afford the 17-ketone **14-3**. Reaction with acetylene in the presence of base, or alternatively lithium acetylide, yields intermediate **14-4**. Attack from the more open backside leads to formation of the α-acetylide. Mild hydrolysis of the enol ether function then affords the unconjugated ketone ethynodrel (**14-5**), This first oral contraceptive to reach the market has become better known as 'The Pill'. The double bond in that product moves into conjugation on treatment with strong acid, forming ethynodrone (**14-6**), the second contraceptive pill to reach pharmacy shelves.

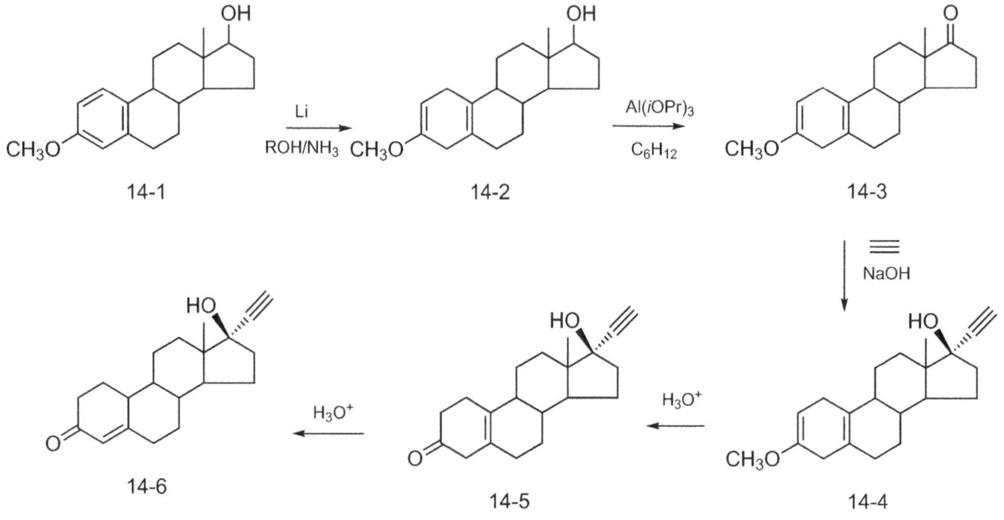

Scheme 4.14

A small portion of **14-1** that fails to react on kilogram-scale Birch reduction will tag along the next set of transformations: oxidation and subsequent ethynylation. This leads to contamination of the final product by the potent estrogen 17-ethynylestradiol 3-methyl ether (**15-2**). This very small impurity in the final product is now known to be essential for contraceptive efficacy. A carefully measured small amount of an estrogen is now included in virtually all oral contraceptives.

In a somewhat related scheme, reaction of estrone methyl ether with lithium acetylide leads to formation of the α-acetylide **15-2** (Scheme 4.15). A small amount of methylamine is then added to a solution of that intermediate in ammonia and an organic solvent.. A sample of lithium metal is added along with ferric nitrate; the latter catalyzes

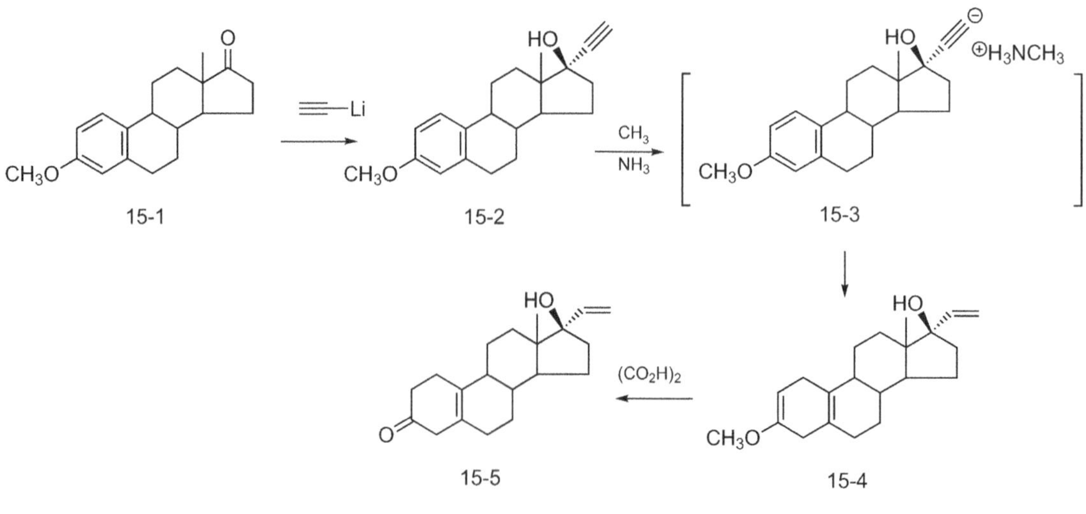

Scheme 4.15

conversion of the metal to the strong base lithium amide. That base then strips a proton from the methylamine. The methylamido anion abstracts the terminal hydrogen on the acetylene **15-3**; the resulting salt decreases the avidity of the triple bond for hydrogen. Addition of additional lithium to the solution then reduces the aromatic ring; concomitant addition of a pair of hydrogens to the acetylene moiety converts the acetylide to a vinyl group (**15-4**). Hydrolysis of the enol ether in that product with oxalic acid gives the progestin norgestrone (**15-5**).

Oxirane **16-1** comprises the starting material for a gonane equipped with an acetonitrile instead of ethynyl side chain at C_{17}. Treatment of the oxirane with potassium cyanide opens the strained three-membered ring to form an acetonitrile side chain (**16-3**) (Scheme 4.16). Exposure to weak acid then hydrolyzes the enol ether at C_3 while keeping the double bond in place. Reaction of that product with bromine gives adduct **16-4** of undefined stereochemistry. Dehydrohalogenation by means of pyridine restores the olefin at C_4 while adding an additional double bond in ring B (C_9–C_{10}). The product, **16-5**, is named dienogest.

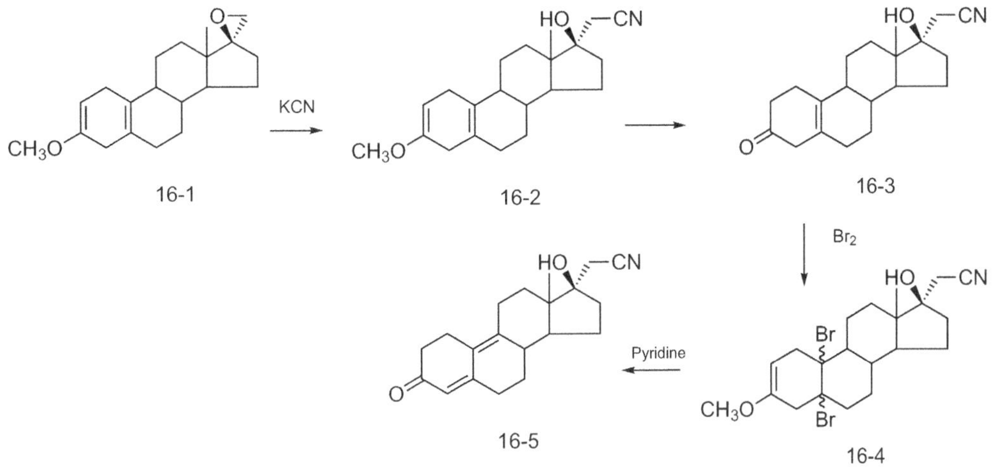

Scheme 4.16

Synthesis of a gonane that bears the acetyl side chain present in progesterone itself (Figure 4.2) starts with the pregnenolone derivative **17-1** obtained from the yam root (*Dioscorea*) process. Oppenauer oxidation of the hydroxyl group a C_3 leads to the diketone **17-2** (Scheme 4.17). Application of the steps involved in the conversion of DHEA to estrone (bromination, dehydrobromination) leads to the 1,4-dien-4-one **17-3**. That product is next aromatized by heating with lithium in the presence of diphenylmethane as buffer to afford the aromatic A-ring intermediate **17-4**. The newly formed phenol is then converted to the methyl ether by means of iodomethane and base. Epoxidation of conjugated olefins is usually carried out by reaction with basic hydrogen peroxide rather than with peroxy acids. Accordingly, reaction of the conjugated olefin in **17-5** with hydrogen peroxide and base leads to the α-epoxide **17-6**.

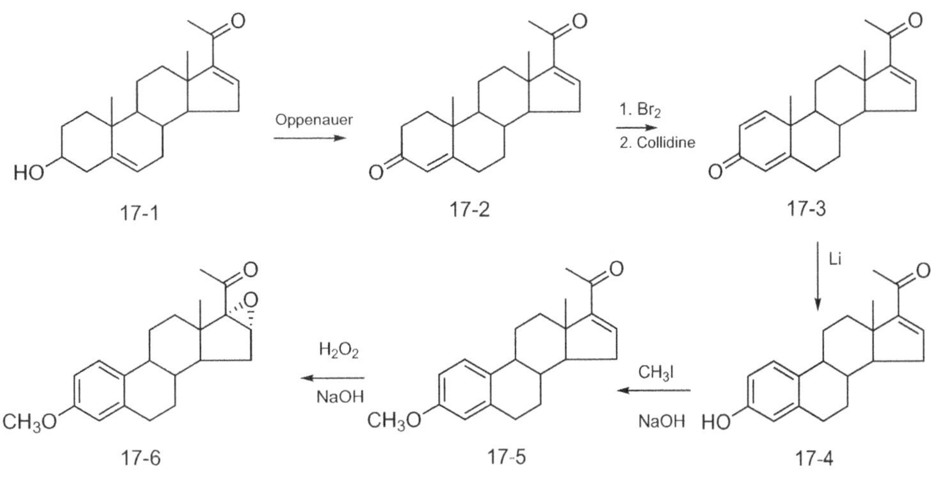

Scheme 4.17

Reaction of the oxirane ring in **17-6** with hydrobromic acid proceeds to yield the bromohydrin **18-1** (Scheme 4.18). Opening of the ring in the other direction is precluded by the diaxial opening preference and also the crowded milieu around C_{17}. The halogen is next removed by treatment with zinc in acetic acid to afford **18-2**. In order to preserve the side chain carbonyl group, that function is converted to its acetal **18-3** by means of ethylene glycol and a small amount of an acid catalyst. Birch reduction of this product converts ring A to a dihydrobenzene (**18-4**). Treatment of this last intermediate cleaves the acetal protecting group and also the enol ether at C_3; the double bond at the ring fusion shifts into conjugation under reaction conditions to give the 19-norprogestin gestonerone (**18-4**).

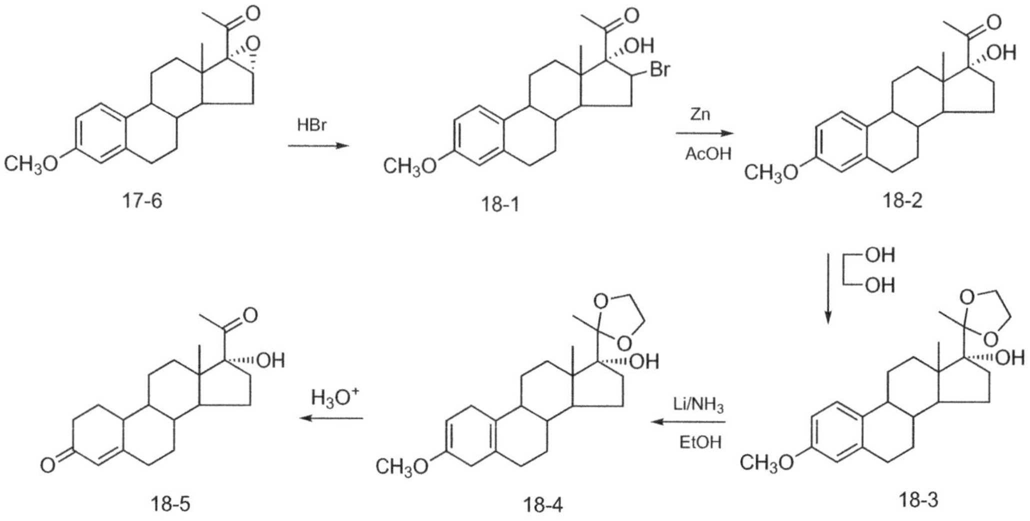

Scheme 4.18

4.3.2.2 Modification at C_7

Addition of a methyl group at C_7 is known to increase the potency of androgens in the 19-methyl series. Adding that group to a gonane invokes the need to go back to testosterone acetate itself (**19-1**) as starting material. Reaction of that compound with chloranil extends the conjugated system by abstracting hydrogen in ring B (**19-2**) (Scheme 4.19). Condensation of the product with methylmagnesium bromide in the presence of cuprous ion adds the required group from the α-side of the terminus of the conjugated dienone system (**19-3**). The acetate protecting group at C_{17} is then saponified. Oxidation of the newly revealed hydroxyl group, for example with Jones' reagent, gives the corresponding ketone, **19-4**. A second dehydrogenation, this time with the more potent oxidizing agent DDQ, establishes the 1,4-dien-3-one system required for expelling the methyl group at C_{10}. Ring A is then aromatized by means of the now standard protocol: reaction with lithium in the presence of diphenylmethane (**19-6**).

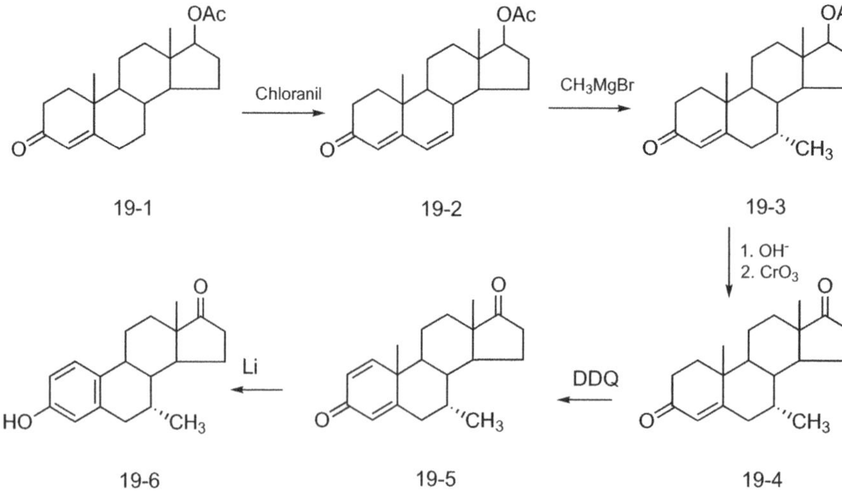

Scheme 4.19

Gonanes or 19-nor-Steroids 59

Alkylation of the phenolic hydroxyl in **19-6**, for example with dimethyl sulfate and base, gives the corresponding methyl ether. The carbonyl group at C_{17} is next reduced by reaction with lithium aluminum hydride (**20-1**) in order to avoid formation of potential mixtures at C_{17} in the subsequent Birch reduction. Thus, treatment of **20-1** with lithium and an alcohol in liquid ammonia leads to the dihydro derivative **20-2** (Scheme 4.20). Re-oxidation of the hydroxyl group in ring D, this time using Oppenauer conditions, leads to the corresponding 17-ketone. Condensation with acetylene and base then gives the adduct **20-3**. Hydrolysis with mild acid leads to scission of the enol ether at C_3. The double bond at the 5(10) position stays in place under those conditions to afford the 19-norprogestin tibolone (**20-4**).

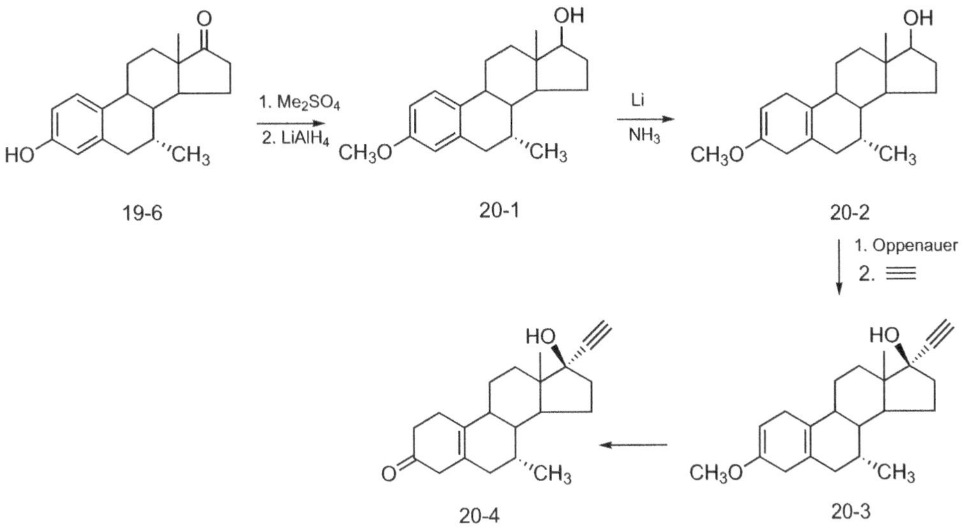

Scheme 4.20

The majority of estrogen antagonists consist of non-steroid compounds. Preparation of one of the rare steroid-based estrogen antagonists invokes a reversal of the usual order of events: gonane chemistry followed by aromatization to an estrane. Reaction of the Grignard reagent from the trimethylsilyl ether of 1-bromononan-9-ol to the gonadienone **19-2** proceeds at the terminus of the 4,6-dien-3-one system to afford the addition product **21-1** as a mixture of epimers at C_7 (Scheme 4.21). The 7α isomer is separated from the mixture before proceeding. The silyl protecting group at the end of the chain is next removed with dilute acid and the newly revealed hydroxyl group is converted to its acetate with acetic anhydride (**21-2**). In a key reaction in the sequence, treatment with cupric bromide removes two hydrogens, aromatizing ring A in the process. Saponification then selectively removes the less hindered hydroxyl group at the end of the long

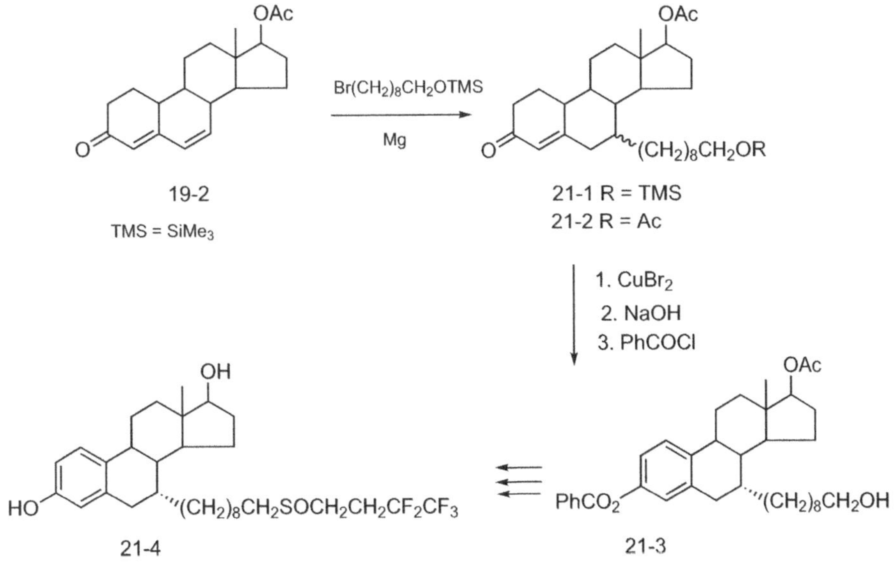

Scheme 4.21

side chain. The phenolic hydroxyl group in the product is acylated with benzoyl chloride to afford the 3-benzoyl intermediate **21-3**. The lengthy sequence continues by converting the hydroxyl to its mesylate and then displacing that good leaving group at the end of the long side chain with 3,4-perfluorobutanethiol; oxidation of sulfur to the sulfoxide affords the estrogen antagonist fluverestant (**21-4**).

4.3.2.3 Modification at C_3

At first sight, it might appear from the structure and biological activity of the progestin lynestrol (**22-6**) that oxygen at C_3 is not necessary for biological activity. It is not unlikely, however, that the deleted oxygen may be re-introduced *in vivo*. The synthesis of that compound starts with Birch reduction of estradiol 3-methyl ether. The dihydro product, **22-2**, is in this case not isolated but is instead treated with ethylenedithiol and a catalytic amount of acid (Scheme 4.22). This procedure leads directly to the dithioacetal **22-3** in which the double bond has shifted to C_4 rather than C_5 as in the case of oxygen acetals. Treatment of the thioacetal with Raney nickel removes the sulfur, in essence converting C_3 to a methylene group. The hydroxyl group in ring D is next oxidized to the corresponding ketone (**22-5**). Adding the ethynyl group with lithium acetylide then gives the progestin lynetrol (**22-6**).

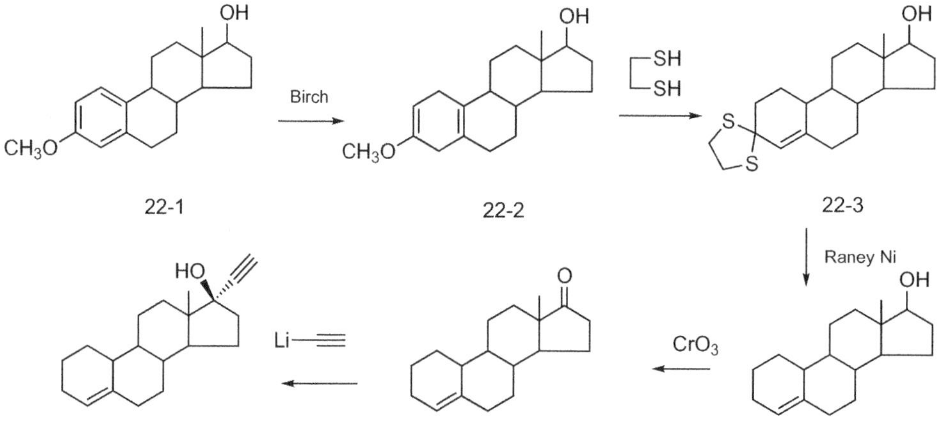

Scheme 4.22

4.3.2.4 Modification at C_{18}

The availability of several total syntheses for gonanes makes it possible to prepare steroids with modifications that could not be obtained from starting material derived from the naturally occurring steroids. A discussion as to whether those modified target molecule present pharmacological advances over compounds with more traditional structure is well beyond the scope of this volume.

Steroids in which the methyl group at C_{18} is replaced by the next higher carbon fragment, ethyl, nicely illustrates this point. Carrying out the first step of the Smith–Torgov synthesis where ethyl replaces the customary methyl group at position 2 in cyclopenta-1,3-dione produces the estrane **23-3**, where the side chain at C_{13} has been extended by one methylene group (Scheme 4.23). That intermediate it then taken on to the gonane by the standard sequence used for the lower homologue, producing the progestin norgestrel (**23-4**). Reaction with acetic anhydride under forcing conditions gives the corresponding 17-acetate **23-5**. Reaction of the acetate with hydroxylamine then affords norgestimate (**23-6**).

4.3.2.5 Additional Unsaturation

As noted previously, anabolic–androgenic activity is enhanced in agents that carry additional double bonds in rings B and C. The same enhancement applies to progestins based on the triply unsaturated 4,10,11-trien-3-one gonane. The higher reactivity towards organometallic reagents of the sterically quite open 3-ketone compared with the hindered milieu of the ketone in ring D in these compounds requires some juggling of protecting groups. Thus, reaction of the 3,17-diketone **24-1**, from oxidation of **12-2**, with the elements of hydrogen cyanide leads to formation of a cyanohydrin at the more electrophilic ketone at C_{17} (Scheme 4.24). Treatment of this intermediate with hydroxylamine gives the C_3 oxime **24-3**. Mild base then reverses the cyanohydrin to the parent ketone to afford the corresponding 3-oximino-17-ketone **24-4**.

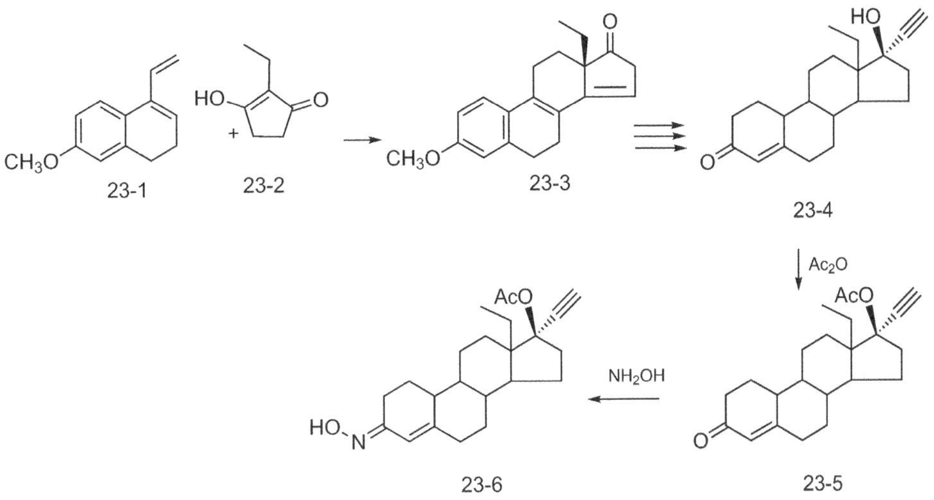

Scheme 4.23

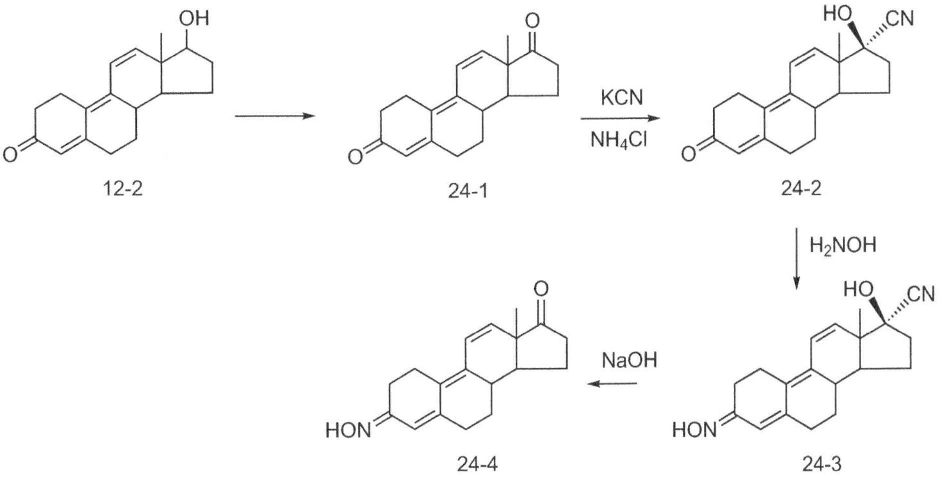

Scheme 4.24

Reaction of the ketone in ring D with lithium acetylide leads to backside addition to give the α-acetylide. Hydrolysis of the oxime with dilute acid then gives the progestin norgestatriene (**25-6**). Alternatively, condensation of **24-4** with allylmagnesium bromide first affords the α-allyl derivative, then hydrolytic removal of the oxime affords the progestin altrenogest (**25-2**) (Scheme 4.25).

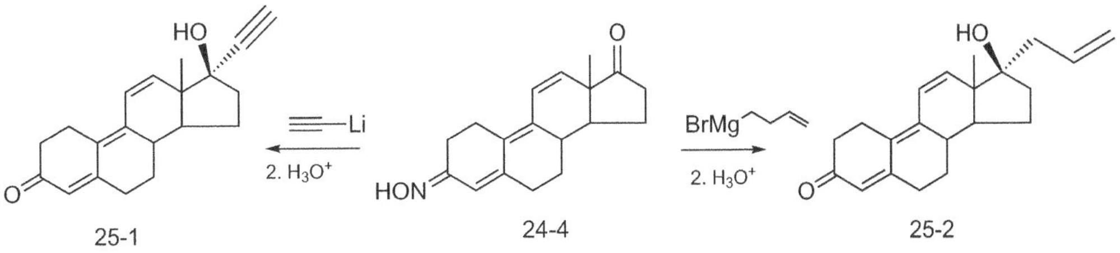

Scheme 4.25

A somewhat different strategy is used for preparing a progestin with additional unsaturation in ring D. The first step in the sequence comprise fermentation of norgestrel intermediate **26-1** with *Penicillium raistricki*. This process inserts an α-hydroxyl group at C_{15} in ring D (**26-2**) (Scheme 4.26). The carbonyl group at C_3 is then converted to its acetal **26-3** by reaction with neopentyl glycol. Hindrance about the D-ring carbonyl insures clean reaction with the C_3 ketone. The double bond in 19-methyl steroids moves to C_5 on conversion of a carbonyl at C_3 to a ketal; in the case at hand, this reaction produces a mixture of C_5 and $C_{5(10)}$ isomers (**26-3**). The total mixture is carried on as such in succeeding steps.

The new hydroxyl group is then converted to a good leaving group, a mesylate (methanesulfonate; **26-4**) by means of methanesulfonyl chloride. Reaction of this intermediate with weakly basic sodium acetate causes elimination of mesyl acid to give the enone **26-5**. Condensation with ethynylmagnesium bromide leads to backside attack and formation of the α-acetylide **26-6**. Hydrolysis of the acetal with aqueous oxalic acid then liberates the carbonyl, to form the progestin gestodene (**26-7**).

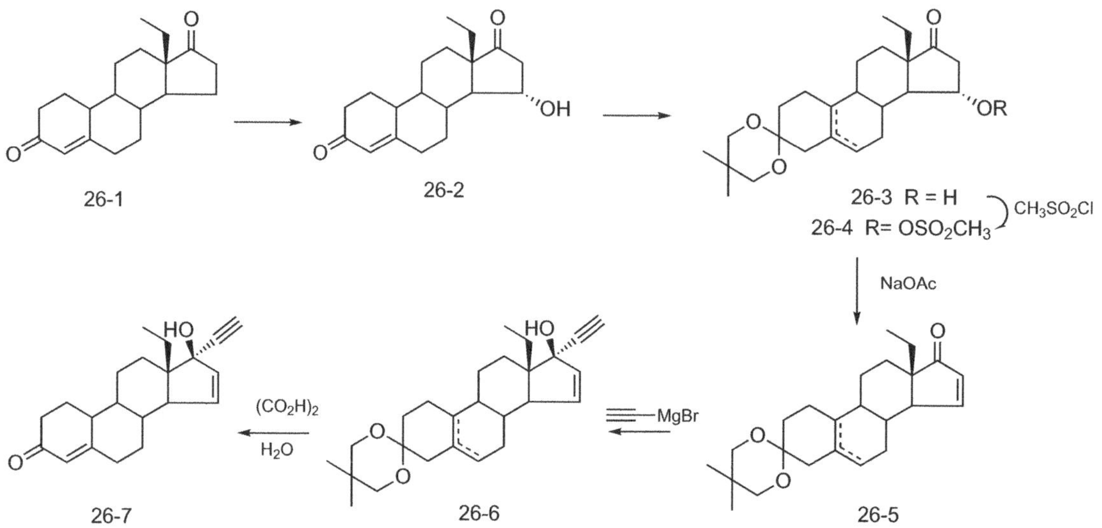

Scheme 4.26

4.3.2.6 Modification at C_{11}

4.3.2.6.1 Methylene Groups
Modifications at C_{11} are notoriously difficult to effect in steroids with a 19-methyl group as the result of its remoteness from functional groups and crowding by the adjacent methyl groups at C_{18} and C_{19}. Although the site is more accessible in the 19-nor series than in, say, pregnanes, relatively long routes are required for preparing compounds modified at C_{11}. One of those schemes begins with a very early intermediate in the norgestrel synthesis (**27-1**) (Scheme 4.27).

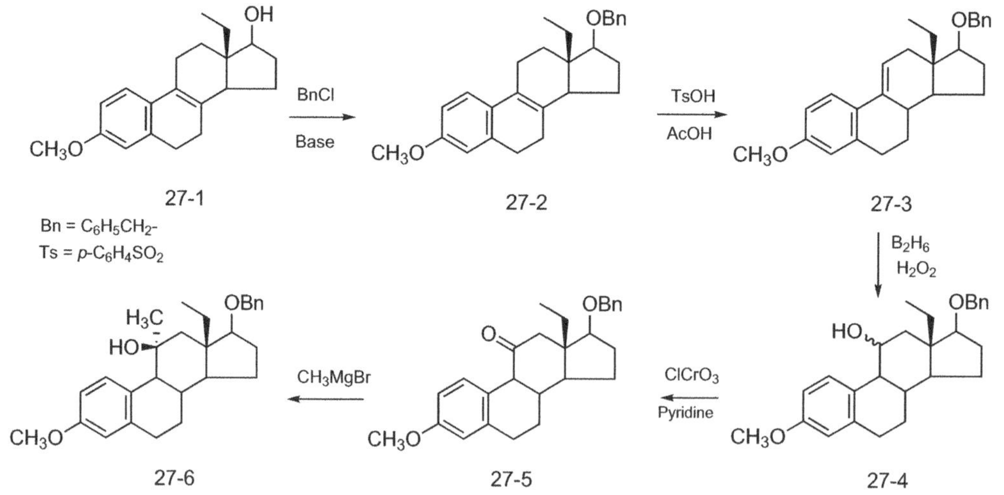

Scheme 4.27

The sequence starts out with the protection the hydroxyl group at C_{17} as its benzyl ether by reaction with benzyl chloride in the presence of base such as sodium carbonate (**27-2**). Treatment of a solution of the benzyl ether in acetic acid with *p*-toluenesulfonic acid causes the styrenoid bond to shift from the B–C ring fusion to the more transoid, and presumably more stable, 9,11-position. This now provides a means for activating C_{11}. Hydroboration with diborane followed by oxidation of the initial adduct with hydrogen peroxide affords the 11-alcohol as a mixture of epimers. Oxidation with chromyl chloride in pyridine leads to the C$_{11}$-ketone **27-5**. Condensation of this product with methylmagnesium bromide proceeds to add from the bottom side to yield the adduct **27-6**. It is of note by way of contrast that the analogous reaction in the normal 19-methyl series proceeds only under the most drastic, forcing conditions and then in low yield.

Reduction of the product of this last reaction, **27-6**, with lithium and *tert*-butanol in liquid ammonia proceeds to the corresponding dihydrobenzene. The *O*-benzyl ether bond is cleaved reductively under those conditions to free the 17-hydroxy group. Hydrolysis with mineral acid then cleaves the methyl enol ether to give the 4-en-3-one function (**28-1**) (Scheme 4.28). Oxidation by means of Jones' reagent converts the hydroxyl at C$_{17}$ to a ketone, protecting oxygen at that position from possible dehydration in the next step. Reaction of **28-2** with *p*-toluenesulfonic and formic acids results in loss of the 11-hydroxy group and formation of the exocyclic methylene (**28-3**); it is of note that the olefin remains as formed instead of moving into the ring. Reaction of that product with pyrrolidine forms an enamine at C$_3$, the less hindered of the two ketones in the compound. The required acetylene side chain is then introduced by reaction of **28-3** with lithium acetylide to give **28-4**. Aqueous acid then hydrolyzes the enamine to afford the progestin etonogestrel (**28-5**).

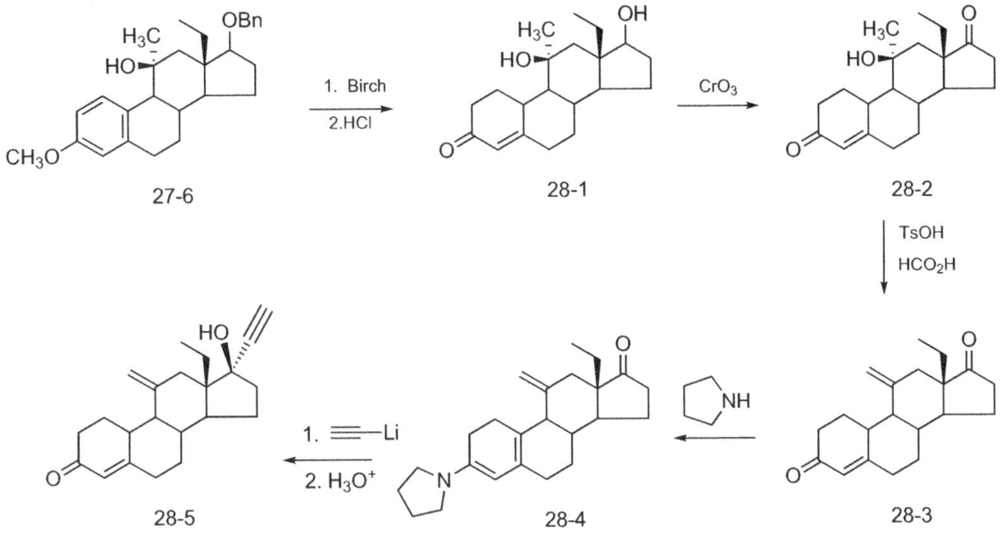

Scheme 4.28

As in the case of lynestrol (**22-6**), pharmacological activity is retained or perhaps even augmented when oxygen at position C$_3$ is deleted. The same holds true in the case of a progestin with an 11-methylene function. Preparation of that compound starts with the etonogestrel intermediate **28-3**. Reaction with ethylene thioglycol leads cleanly to he thioketal **29-1** (Scheme 4.29). The usual selectivity for acetal formation at C$_3$ rather than C$_{17}$ is in this case enhanced by the bulk

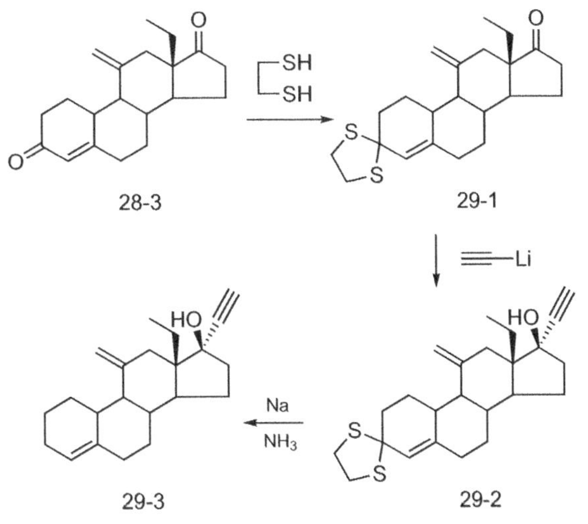

Scheme 4.29

of the additional methylene group at the latter position. Addition of lithium acetylide affords the ethynylated intermediate **29-2**. Reduction of the thioketal, in this case with sodium in liquid ammonia, removes sulfur from C_3 to afford desogestrel (**29-3**). The proton on the free end of acetylide moiety is probably removed under the reaction conditions to generate a negative charge on that carbon atom. This likely accounts for the survival of the acetylene.

4.3.2.6.2 Aryl Groups Increasing the size of the group at C_{11} from methylene to benzene drastically alters the pharmacological properties of gonanes. The former, as noted above, generally bind to progestin receptor sites where they act as agonists. The 11-aryl compounds, on the other hand, generally bind to progestin and also glucocorticoid sites. These 11-arylgonenes act as antagonists at those sites. Selectivity for progestin and/or glucocorticoid sites is determined by the substituents on the benzene ring and, to some extent, the substitution on C_{17}.

The synthesis of the best-known member of this series, mifepristone (RU-486), starts with an intermediate, **5-4**, from the total synthesis by successive ring closures. Saponification of intermediate **5-4** removes the benzoate protecting group in ring D. The resulting alcohol is oxidized to the corresponding ketone. The ketone at C_3 is then protected as its acetal, **30-1** (Scheme 4.30). Reaction of this intermediate with trimethylsiliyl cyanide leads to addition to the ketone at C_{17} to form the silyl ether of a cyanohydrin (**30-2**), a protecting group that is very stable to base. Treatment of this intermediate with the adduct from hydrogen peroxide and perfluoroacetone selectively forms an α-epoxide at the A–B ring fusion (**30-3**). Selectivity over the olefin at position 9(11) may be due to steering action by an oxygen atom on the C_3 acetal. Reaction with the Grignard reagent from 4-dimethylaminobromobenzene can be viewed as a vinylogous oxirane ring opening (**30-4**). The sequence may be rationalized by assuming attack by the Grignard reagent on the olefin at C_{11}, concomitant shift of the olefin to C_9 and opening of the oxirane. The net result, formation of the β-aryl-5α-hydroxy alcohol **30-5**, follows the rule of diaxial oxirane ring opening.

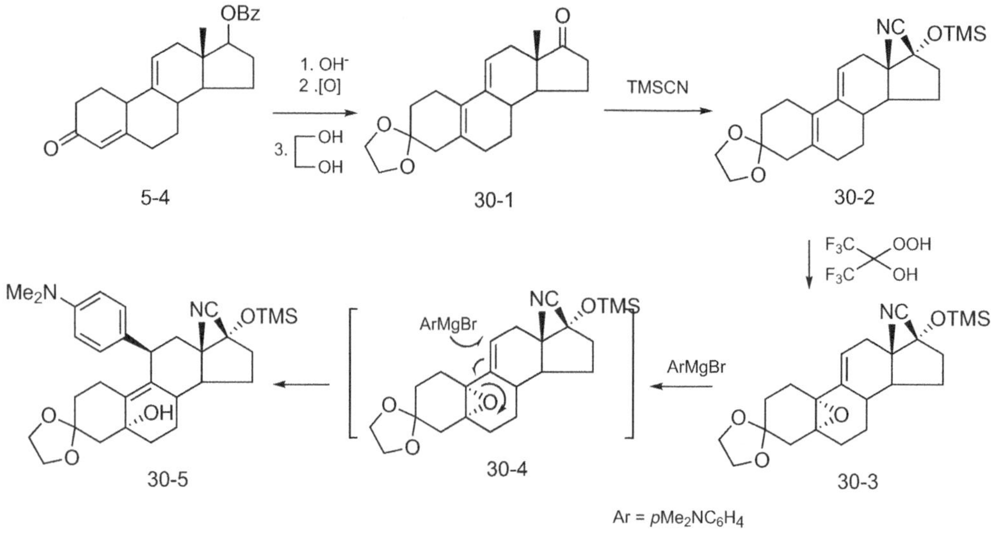

Scheme 4.30

Hydrolysis with mild acid leads to selective reversal of the silyl cyanohydrin formation, thus revealing the ketone at C_{17}. Condensation of that function with lithium acetylide proceeds as in all cases noted earlier to give the 17α-ethynyl-17β-hydroxy intermediate **31-2** (Scheme 4.31). Treatment of that intermediate with acid then leads to hydrolysis of the ketal at position 3. The transient intermediate hydroxy ketone quickly loses water so as to form an olefin at C_4 and thus mifepristone (**31-3**).

The order of some of the steps is reversed in the synthesis of an 11-aryl agent that bears an oximinocarbonyl group on the aromatic ring. The scheme starts in much the same way as the preceding one. The first step involves protecting the carbonyl function at C_3, in this case as an open-chain bismethoxy acetal (**32-2**) (Scheme 4.32). Treatment with hydrogen peroxide then converts the ring fusion olefin to an α-oxirane (**32-3**) Condensation with the Grignard reagent from the methoxyacetal of 4-bromophenylcarboxaldehyde (**32-4**) again leads to vinylogous opening of the epoxide to afford the adduct **32-5**. Building the side chain at C_{17} differs significantly from the previous example. Thus, reaction of **32-5** with the carbanion from trimethylsulfoxonium iodide and strong base inserts a methylene group into the C=O bond of the ketone at C_{17} to form a new oxirane (**32-5**), with the additional carbon atom entering from the backside. Sodium methoxide then opens the epoxide to give the α-methoxymethyl derivative **32-7**.

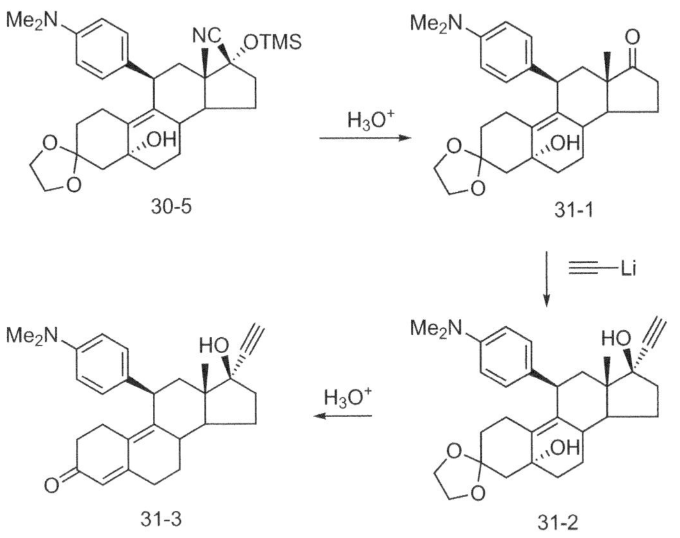

Scheme 4.31

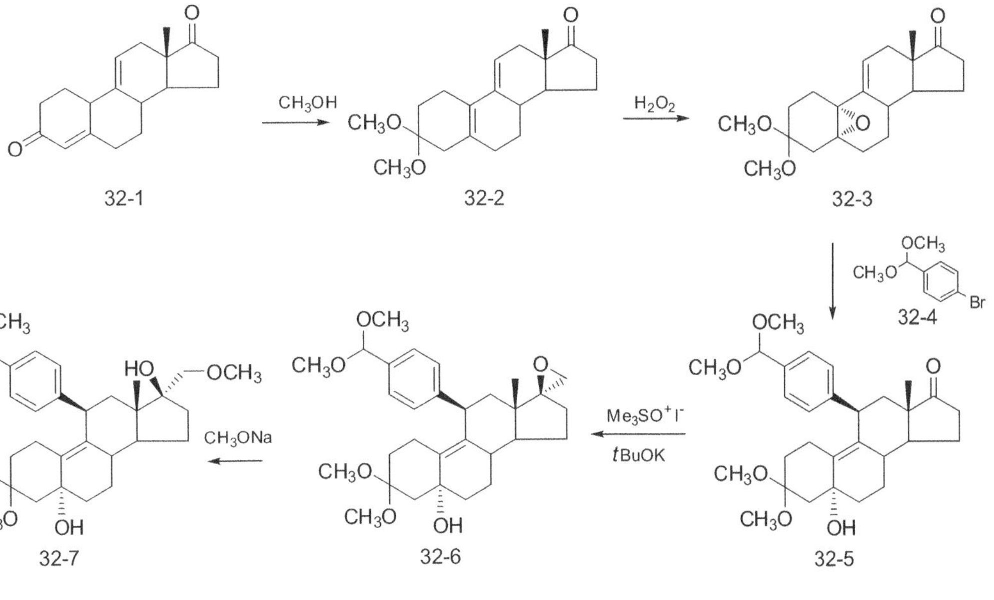

Scheme 4.32

Acid hydrolysis of **32-7** leads to hydrolysis of both acetal groups in **32–7** as well as dehydration of the C_5 hydroxyl group to give the penultimate intermediate **33-1** (Scheme 4.33). Reaction of the aldehyde on the aromatic ring with hydroxylamine affords the oxime **33-2** and thus asoprinsil.

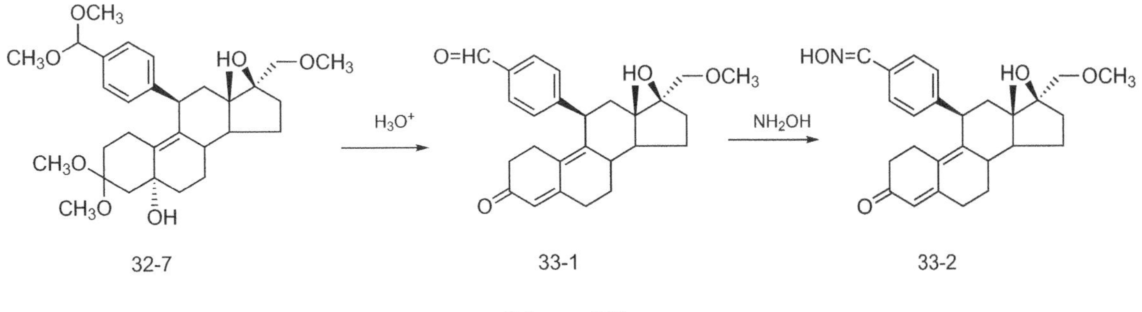

Scheme 4.33

The starting material, **34-1**, comprises the neopentyl glycol-protected analogue of the previous 5(10)-epoxides (**30-3**, **32-3**). Condensation of that oxirane with the Grignard reagent from neopentyl glycol acetal of 4-bromoacetophenone,

34-2, leads to the 11-aryl-substituted intermediate **34-3** (Scheme 4.34). The presence of a perfluoroethane side chain on C_{17} is the distinguishing feature that sets this agent, **34-5**, apart from the previous examples. This feature is incorporated by condensing **34-3** with the lithio reagent generated from metal interchange between pentafluoroethyl iodide and methyllithium (**34-4**). Exposure to strong acid strips off both acetal groups and causes dehydration of the β-hydroxy ketone. The product **34-5** is described as a pure progesterone antagonist with little activity at glucocorticoid receptors.

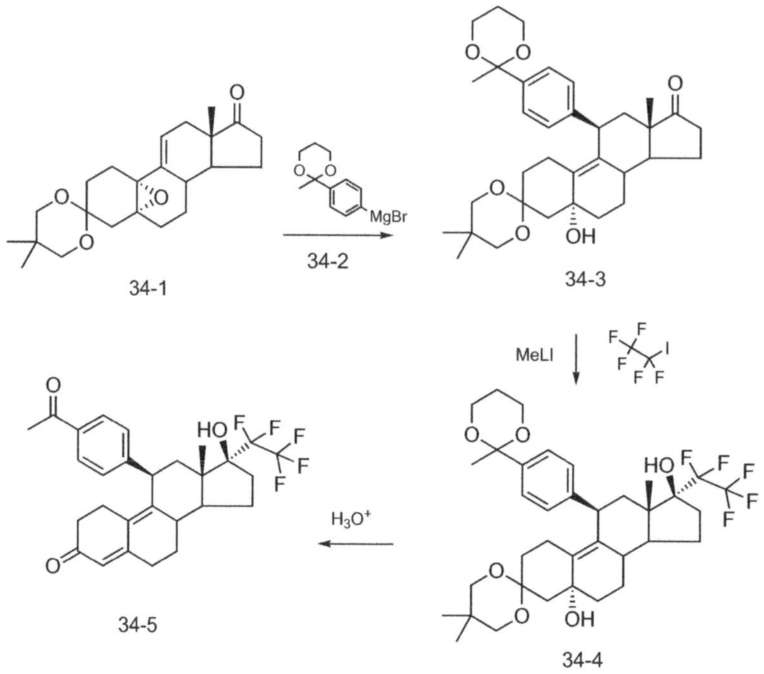

Scheme 4.34

Cis stereochemistry at the fusion between the five- and six-membered rings comprises the preferred configuration for hydrindanes. The *trans* configuration between rings C and D that maintains in all naturally occurring steroids and their derivatives actually represents the energetically disfavored ring fusion configuration. Photolysis of the advanced 11-aryl intermediate **35-1** for 11-arylgonane syntheses inverts the C–D fusion from *trans*- to *cis*-**35-2**. The reaction can be rationalized by assuming that photolysis first cleaves the C_{13}–C_{14} bond at the ring fusion to yield a transient diradical such as **35-2** (Scheme 4.35). That reactive species then closes to form the energetically favored *cis* isomer (**35-3**). The

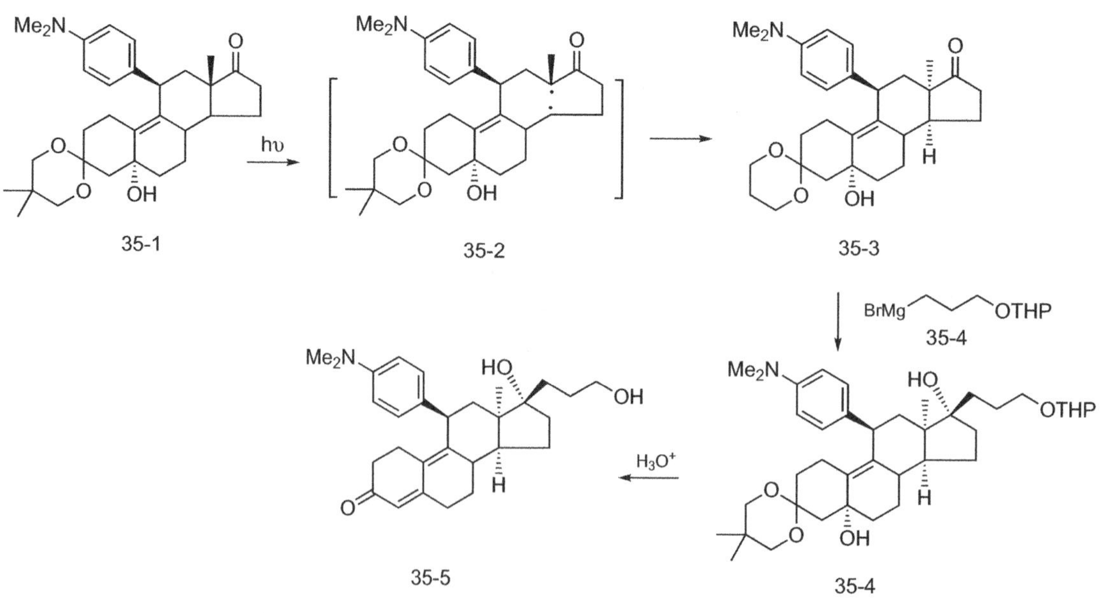

Scheme 4.35

altered milieu about C_{17} that includes an adjacent α-methyl group inverts the preferred approach for additions to the ketone at C_{17}. The Grignard reagent from tetrahydropyranyl (THP)-protected 3-bromopropanol adds to give a product that consists of a β-alkyl-α-hydroxy isomer (**35-4**). Acid hydrolysis removes all protecting groups to yield the progestin blocker onapristone (**35-5**).

4.4 Some Drugs Based on Gonanes

4.4.1 Androgenic–Anabolic Agents

The anabolic action of androgens, as noted earlier, has made them particularly attractive to body-builders and athletes. This activity is also, it can be speculated, responsible for referring to any heightened response by the phrase '... on steroids'. Only one of the small group of androgens discussed above, norethandrolone (**6-3**), is currently marketed as a prescription androgen–anabolic agent. This drug was approved for sale by the FDA in 1965 and finds legitimate use as an androgen for cases of hormonal deficiency. The other action of this and other androgens, anabolic activity, helps to maintain nitrogen balance in the body. This leads to accumulation of new protein in muscle tissues. Formal indications that arise from the anabolic activity include uses such as treatment of patients with severe burns, after severe trauma and for certain rare forms of aplastic anemia. An attractive property of norethandrolone is that it is orally active. The potential competitor, norbolethone (**7-4**), was never marketed due to concern about the drug's safety in spite of its proven activity in human trials.

Mibolerone (**9-6**) is approved for preventing estrus in female dogs. Significant amounts of this orally active anabolic agent are diverted for use by body builders and athletes. This use is specifically proscribed by the DEA.

Trenbolone (**12-2**) is another androgenic–anabolic agent approved for veterinary use. The drug is administered to cattle during shipment in order to minimize weight loss causes by stress.

4.4.2 Progestins

Norethynodrel (**14-5**) (Enovid), approved for sale in 1962, was the first of the orally active contraceptive progestins. The availability of this class of drugs has had a profound effect on sexual mores in the countries where they have been approved for sale. Launch of norethynodrel preceded the development of many of the current analytical tools for detecting trace impurities, such as thin-layer chromatography and high-performance liquid chromatography (HPLC). The discovery that a very pure sample of a closely related progestin was not nearly as active as the marketed drug led to the finding that the latter was contaminated with a very small amount of the 3-methyl ether of 17α-ethynylestradiol that arose from incomplete Birch reduction. As a result, Enovid was reformulated to include a measured amount of that potent estrogen. Many of the progestins described above are available as oral contraceptives in some parts of the world. These are almost invariably formulated with a small measured amount of an estrogen.

High doses of the gonane-based progestins will prevent incipient pregnancy even when ingested by the woman for some time after intercourse that may have resulted in fertilization. 'Plan B', billed as an over-the-counter emergency contraceptive, consists of a 1.5 mg dose of levonorgestrel, the levorotatory isomer of norgestrel (**24-3**).

4.4.3 Progestin Antagonists

Ovarian follicles, as noted above, secrete increasing amounts of progesterone as the menstrual cycle progresses. The presence of that hormone leads to the development of special tissues on the inner wall of the uterus that are slated to provide a locus for a fertilized ovum. A progesterone antagonist such as mifepristone(**31-3**) (Mifeprex, RU-486) causes those tissues to collapse and as a result make the uterus unreceptive. This leads to expulsion of the fertilized ovum, in effect a chemical abortion. In practice, the drug is administered along with a prostaglandin (commonly misoprostol) to facilitate expulsion of the ovum and the uterine lining. In order to sidestep the furor caused by potential introduction of RU-486, the sponsor, then Roussel-UCLAF, turned the drug over to a not-for-profit corporation.

5
Androstanes, C_{19} Steroids and Their Derivatives

5.1 Biological Activity

Androstanes, and specifically testosterone and its activated derivative 5α-dihydrotestosterone, exercise a role in males in some ways analogous to that which estrogens play in females. The pharmacological activity of the circulating androgens and their modified derivatives is, in most respects, the same as that of the 19-norandrostanes described in the previous chapter. The anabolic activity of compounds in this class causes them to be eagerly sought by body builders and athletes; the C_{19} androgens, as a result, make up the majority of the steroids on the DEA List of Controlled Substances. Note that the configuration of the proton at C_5 (when present) in all the following schemes is α when not specified otherwise.

A small set of steroids prepared from androstanes feature a spirobutyrolactone at C_{17}. These agents act as antagonists of aldosterone, the highly oxygenated steroid that controls serum electrolytes and blood volume. The antagonist action of the spirolactones is manifested as diuretic and antihypertensive activity.

5.2 Sources of Androstanes

5.2.1 From Pregnenolone

16-Dehydropregnenolone (**1-1**), obtained from diosgenin (see Chapter 2, Scheme 2.8), is a key intermediate in the preparation of C_{19} steroids. Reaction of that pregnenolone with hydroxylamine affords the corresponding oxime **1-2** (Scheme 5.1). Treatment of **1-2** with strong acid leads to Beckmann rearrangement and formation of the C_{17} acylenamine **1-3**. Hydrolysis of that product removes the acyl group on both the enamine and the hydroxyl group at C_3. Further reaction hydrolyzes the enamine at C_{17} to a ketone via the corresponding enolate (**1-4**). Oxidation by any of several methods gives androstene-3,17-dione (**1-6**), after the double bond shifts into conjugation.

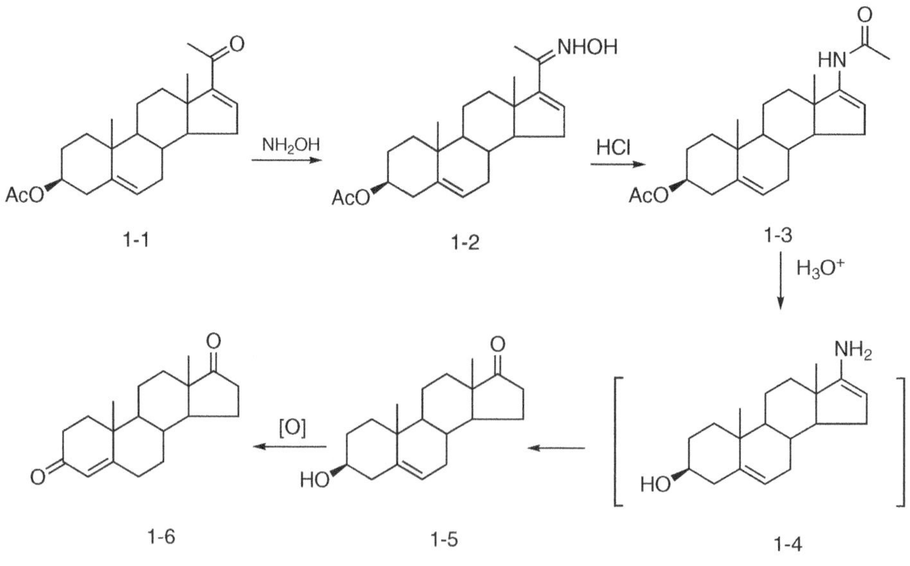

Scheme 5.1

5.2.2 Fermentation

Fermentation of a mixture of soy sterols, sitosterol, stigmasterol and traces of related compounds with *Microbacteria* sp. produces androstenedione (**1-6**). Relatively recent patents describe the production of the steroid from phytoserols from various crude vegetable oils, bypassing the need for prior isolation of the phytosterol fraction.

5.2.3 Total Synthesis

The driving force behind the development of total syntheses for estrane and to some extent gonanes described in Chapters 2 and 3 lay in the then scarce and hence expensive steroid starting materials. The schemes that were developed made possible the elaboration of derivatives not accessible from estrone, such as gonanes with an additional carbon on the angular methyl at C_{13}. Both androstene-17-dione and testosterone have been prepared by total synthesis. The schemes by which that was accomplished, however, were lengthy and complex. Those syntheses mainly represented a *tour de force* for chemical synthesis since they were not competitive with sources of androstanes from pregnenolone or by fermentation.

An eletrocyclization provides the key reaction in one total synthesis of an androstane. It takes approximately a dozen steps to prepare the substrate, **2-1** (shown in 3D in **2-2**) (Scheme 5.2). Heating at 200 °C closes rings A and B to afford the steroid skeleton in a concerted process (**2-3**). Reaction of that product with singlet oxygen, formed for example by irradiation in the presence of a photosensitizer dye, leads to attack at C_3 with concomitant shift of the double bond (**2-4**). Several more steps then give androstenedione (**1-6**).

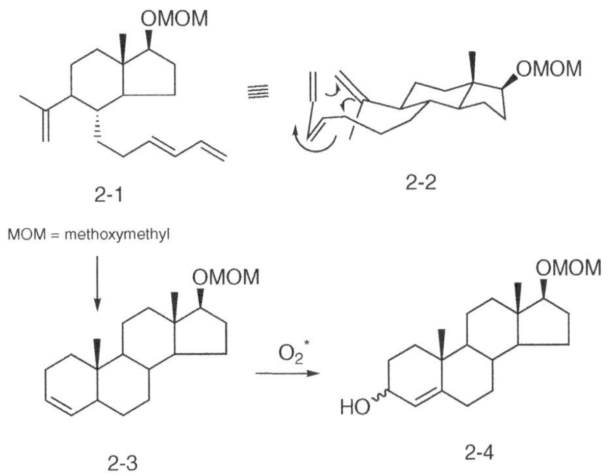

Scheme 5.2

5.3 Modified Anabolic–Androgenic Androstanes

5.3.1 17-Desalkyl Compounds

5.3.1.1 Modification on Ring A

5.3.1.1.1 Oxidation State Preparation of testosterone from the abundant starting compound dehydroepiandrosterone (**1-5**) in essence involves interchanging oxidation states between C_3 and C_{17}, a process that would involve juggling protecting groups. The selectivity with which manganese dioxide oxidizes allylic hydroxyl groups simplifies that transformation.

Thus, reduction of androst-5-ene-1,17-dione (**1-5**) with lithium aluminum hydride reduced both 3- and 17-carbonyl groups, that at C_{17} cleanly to the β-epimer and the ketone at C_3 to a mixture of epimers (**3-1**) (Scheme 5.3). Treatment of the mixture of epimers with manganese dioxide oxidizes the allylic alcohol at C_3, leaving that at C_{17} untouched, affording testosterone (**3-2**).

Reaction of testosterone with DDQ series subtracts hydrogen in ring A, as in the gonane series, to form the cross-conjugated 1,4-diene **4-1** (Scheme 5.4). Acetal exchange of this product with the dimethyl acetal (**4-2**) from cyclopentanone gives the mixed acetal **4-3**. Heating that product in an inert solvent splits out the elements of methanol to give the enol ether quinbolone (**4-4**).

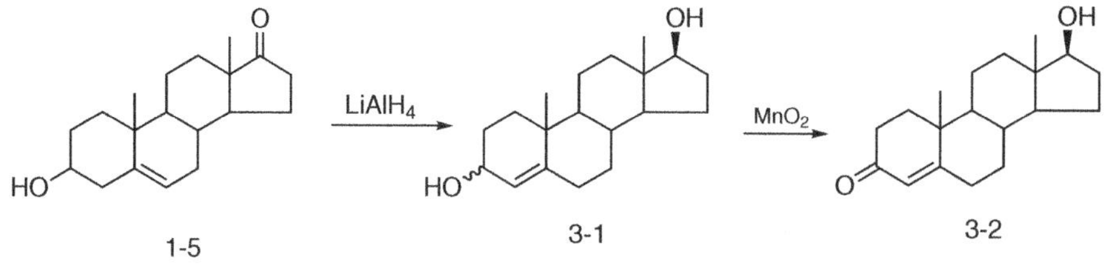

Scheme 5.3

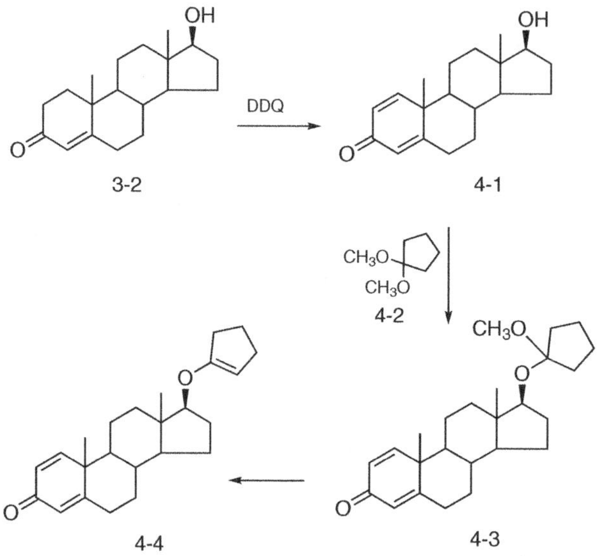

Scheme 5.4

Oxidation of testosterone with osmium tetraoxide gives the corresponding 4,5-glycol (**5-1**) of undefined stereochemistry. By the method of formation, both hydroxyls probably have the same β-configuration. This product the undergoes spontaneous β-dehydration to the enol hydroxide derivative formestane (**5-2**) (Scheme 5.5).

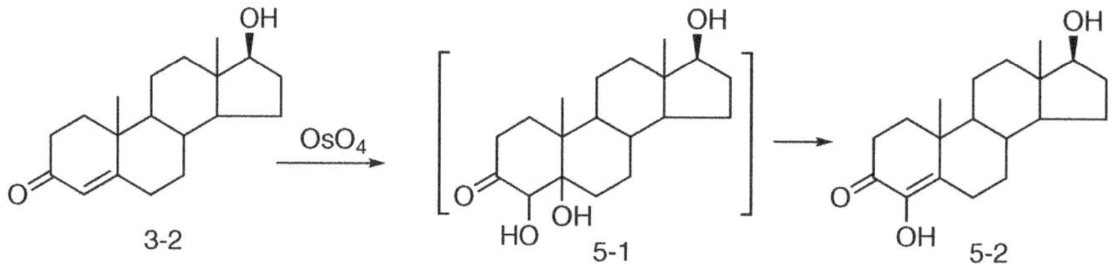

Scheme 5.5

5.3.1.1.2 Additional Carbon Atoms

Acylation of testosterone with propionyl chloride leads to the corresponding 17-propionyl derivative **6-1** (Scheme 5.6). Catalytic hydrogenation proceeds from the more open α side of the molecule to afford the dihydrotestosterone derivative **6-2**. Condensation of that intermediate with ethyl formate and sodium ethoxide proceeds regiospecifically at C_2 to give the formyl derivative **6-3**. Preference for formation of the intermediate enolate at C_2–C_3 instead of C_3–C_4 can be rationalized by the lower internal strain associated with the former. A second hydrogenation step saturates the formyl double bond in **6-3** and at the same time removes the hydroxyl group to give the 2β-methyl derivative (**6-4**).

As a result of backside addition of hydrogen, the newly produced methyl group occupies the energetically disfavored axial 2β-methyl group. Treatment with base inverts that center to give the equatorial 2α-methyl derivative. The propionyl ester is cleaved in the process to the androgen dromostanolone (**6-5**).

In much the same vein, reaction of dihydrotestosterone acetate (**7-1**) with ethyl formate and sodium ethoxide gives the 2-formyl derivative **7-2** (Scheme 5.7). Treatment of this product with DDQ results in the abstraction of hydrogen from C_{1-2} and thus formation of the cross-conjugated derivative **7-3**.

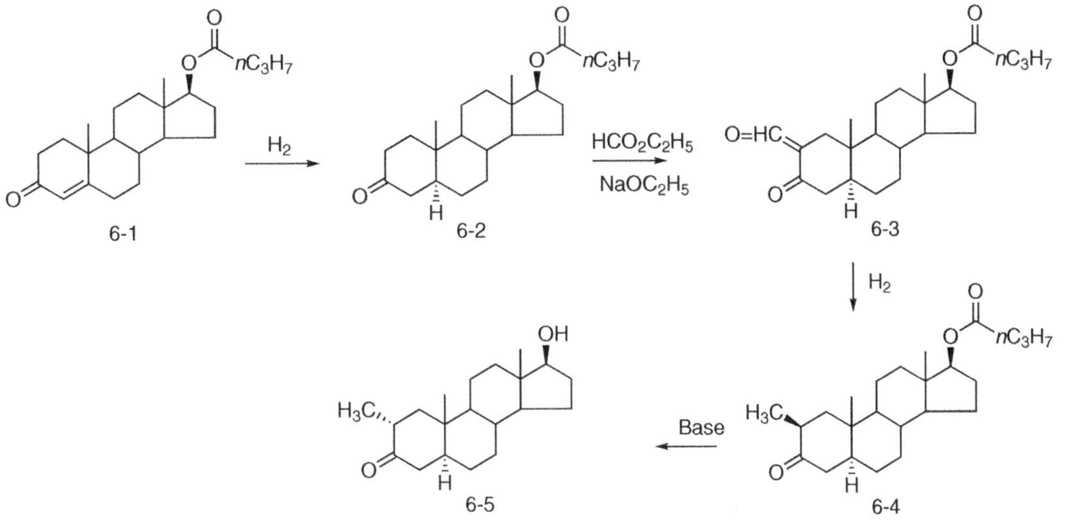

Scheme 5.6

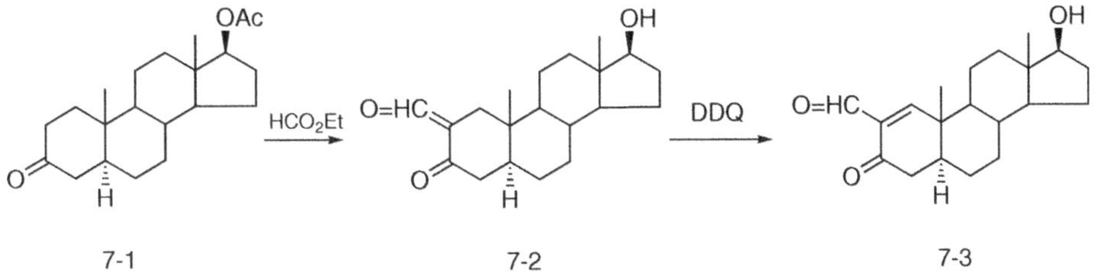

Scheme 5.7

The Mannich reaction, which comprises one of the other methods for grafting a carbon atom on to an enolizable position, is also applicable to steroids. Thus, reaction of dihydrotestosterone acetate (**7-1**) with formaldehyde and dimethylamine affords the 2-dimethylaminomethyl derivative **8-1** (Scheme 5.8). The observed regiochemistry again illustrates the preferred enolization towards C_2. Catalytic hydrogenation at elevated temperature replaces the dimethylamino group by hydrogen. This reaction may actually proceed by reduction of a 2-methylene intermediate from prior elimination of dimethylamine. Bromination then provides the 2-bromo derivative **8-3**. Treatment of this intermediate with lithium carbonate in dimethylformamide (DMF) leads to elimination of the elements of hydrobromic acid and formation of a 2-en-3-one system (**8-4**). This anabolic androgen stenbolone ranks high on the DEA List of Controlled Substances.

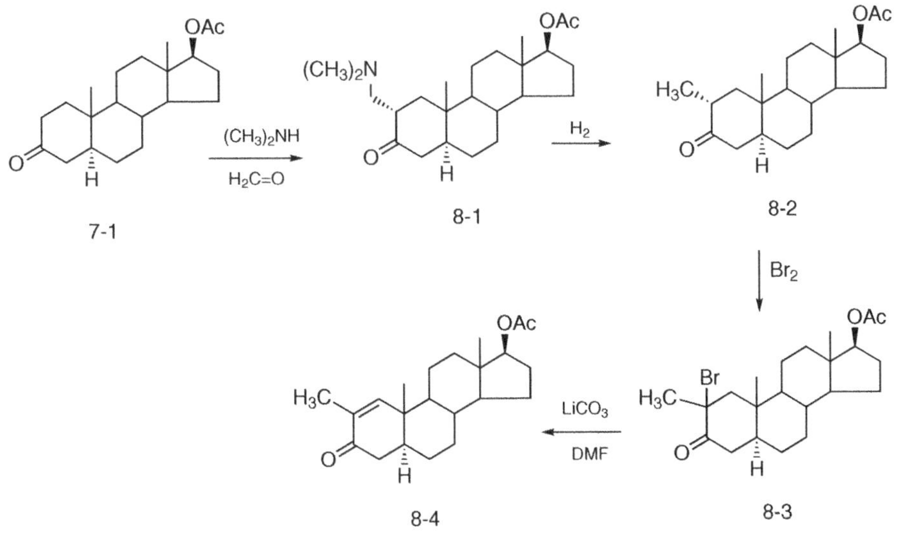

Scheme 5.8

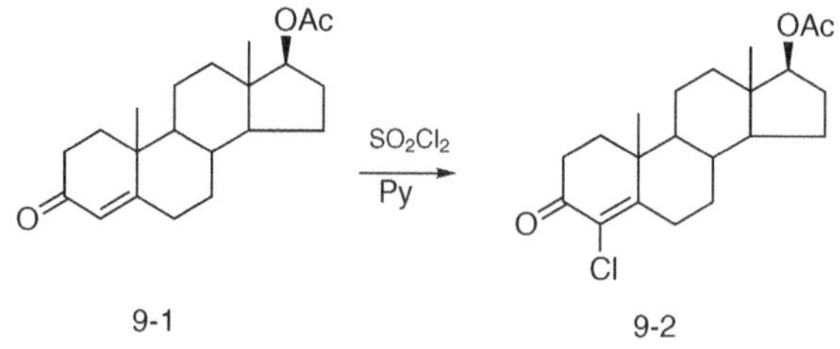

72 *Steroid Chemistry at a Glance*

Scheme 5.9

5.3.1.1.3 Halogenated Derivatives Slow addition of sulfuryl chloride to a cold solution of testosterone acetate (**9-1**) and pyridine results in formation of the 4-chloro derivative **9-2** (Scheme 5.9). It is likely that the reaction, like many other sulfuryl chloride halogenations, proceeds via a free radical mechanism.

Catalytic hydrogenation of androst-4-ene-3,17-dione (**1-6**) proceeds to afford the 5α-dihydro derivative **10-1** (Scheme 5.10). Reaction of this intermediate with potassium perchlorate and sulfuric acid, in essence perchloric acid, leads to formation of the 2α-chloro derivative **10-2**. The regiochemistry may be due to factors similar to those that guide the formylation reactions. Stereochemistry in this case likely reflects preference for the less crowded equatorial epimer.

Incorporating fluorine in the structure of drugs often leads to increased potency over the compound lacking that halogen. Many methods have consequently been developed for adding fluorine to potential drugs. In one such scheme, the formyl derivative **11-1** from testosterone is first treated with sodium methoxide to afford an ambident carbanion such as **11-2** (Scheme 5.11). Subsequent addition of fluorine perchlorate leads to transfer of fluorine from the reagent to the center of charge at C_2. Heating the unstable β-dicarbonyl intermediate **11-3** in the presence of potassium acetate leads to extrusion of carbon monoxide. The weak base present in the reaction mixture allows the halogen to equilibrate between its two configurations. The resulting stereochemistry of the fluorinated product **11-4** again reflects the preferred configuration.

Enol ethers offer another means for incorporating fluorine. Reaction of dihydrotestosterone acetate (**7-1**) with ethyl orthoformate in the presence of a catalytic amount of acid gives the corresponding enol ether **12-1** (Scheme 5.12), again

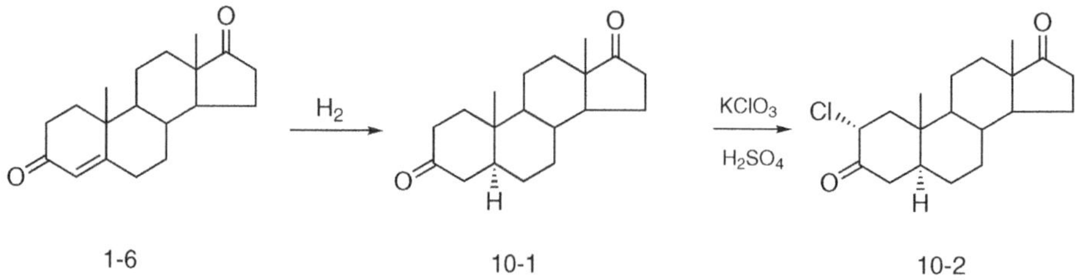

Scheme 5.10

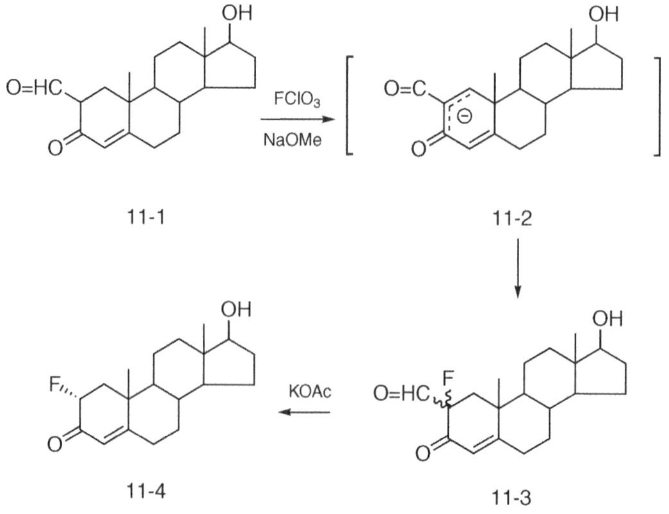

Scheme 5.11

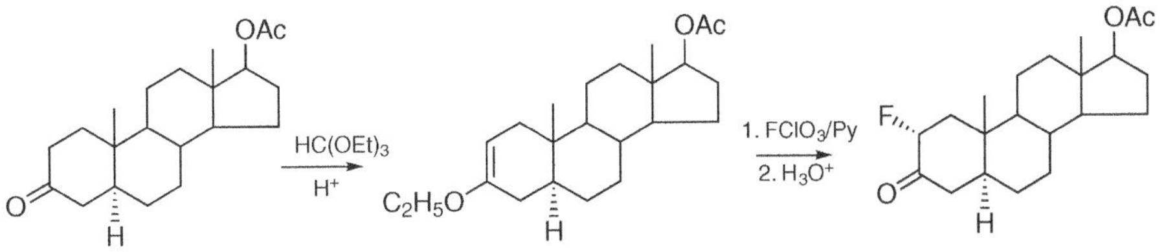

illustrating the preferred enolization towards C_2. Treatment of that product with fluorine perchlorate followed by acid hydrolysis of the enol ether leads to 2-fluorodihydrotestosterone acetate (**12-2**). This procedure also produces the equatorial α-fluoro derivative.

5.3.2 17-Alkyl Compounds

Orally administered testosterone, and also derivatives that feature a secondary hydroxyl group at C_{17}, are rapidly deactivated in the liver. These agents, as a result, never attain therapeutically effective blood levels. This rapid inactivation can be avoided by adding an alkyl group at that position. The chemistry used for such additions generates a 17β-hydroxyl analogous to that present in testosterone. Much of the research on androstanes has as a result been focused on molecules that possess an alkyl group at C_{17}; that substituent is most commonly methyl.

The prototypical 17-alkylandrogen, 17-methyltestosterone, is prepared from dehydroepiandrosterone (DHEA; **1-5**) in two steps. Reaction of DHEA with an excess of methylmagnesium bromide gives the corresponding 17-methyl derivative (**13-1**) (Scheme 5.13). Oppenauer oxidation then converts the hydroxyl at C_3 to a carbonyl group; the olefin shifts into conjugation in the process to give mestanolone (**13-2**).

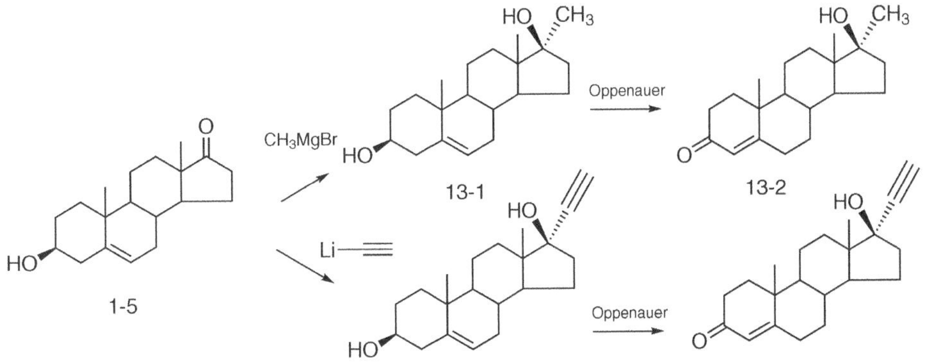

Scheme 5.13

Essentially the same procedure but substituting lithium acetylide for the Grignard reagent gives the 17-ethynyl derivative **13-3**. Oxidation of the hydroxyl group then affords ethisterone (**13-4**), a compound that, perhaps surprisingly, shows some progestational rather than androgenic activity. This finding arguably provided the incentive for preparing the corresponding ethynyl derivatives in the 19-nor series, leading to the first oral contraceptives.

5.3.2.1 Modifications on Ring A

5.3.2.1.1 Alkyl Derivatives Many of the simpler modifications on ring A follow closely the analogous changes in the 17-desalkyl series. Many of those transformations were actually first applied to the 17-alkyl compounds.

Treatment of mestanolone (**14-1**) with selenium dioxide, as in the case of the compound lacking an alkyl group at C_{17}, removes hydrogen from ring A to form a new double bond at C_1, yielding methandrostenolone (**14-2**) (Scheme 5.14).

Catalytic hydrogenation of 17-methylated DHEA (**13-1**) provides the corresponding derivative **15-1**, where hydrogen has been added from the backside (Scheme 5.15). Oxidation by any one of several methods, such as the Oppenauer reaction, restores the ketone at C_3. This compound, **15-2**, comprises starting material for several ring A-modified derivatives.

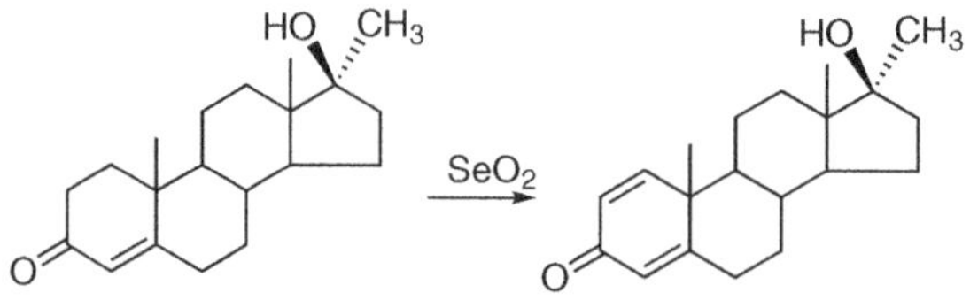

Scheme 5.14

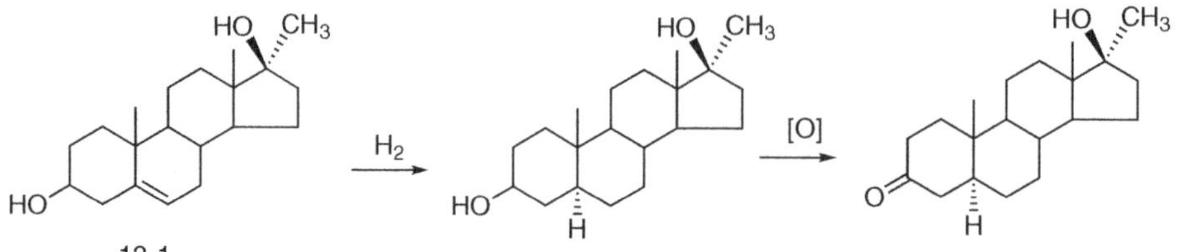

Scheme 5.15

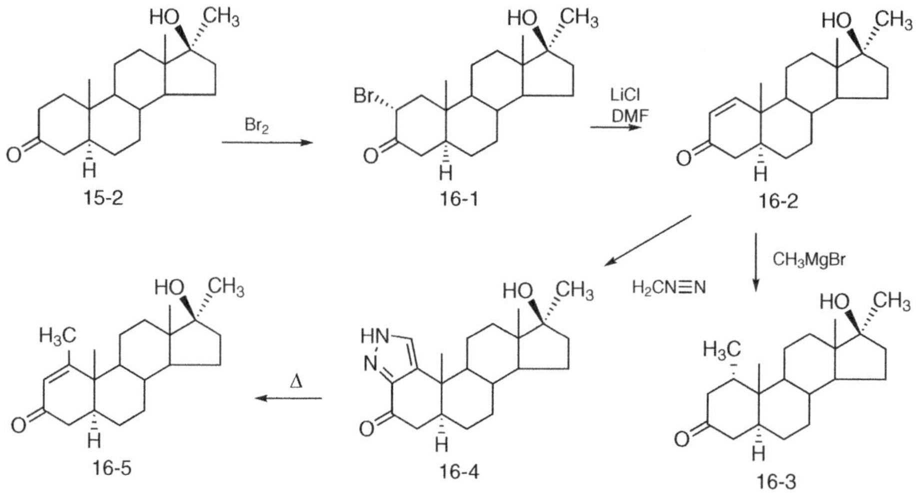

Scheme 5.16

Reaction of that fully saturated derivative (**15-2**) with a carefully measured amount of bromine proceeds to form the 2-bromo derivative **16-1** (Scheme 5.16). Halogen occupies the favored α-position, which in this case is equatorial. Treatment of this compound with lithium chloride in the presence of DMF leads to dehydrohalogenation and formation of the unusual 1-en-3-one compound mesterolone (**16-2**). Methylmagnesium bromide adds to the end of this enone system to form the 1α-methyl derivative. The configuration at C_1 happens to represent the energetically favored equatorial position. This is likely dictated by approach of the reagent from the backside and away from the adjacent angular methyl group; note that the newly added alkyl cannot equilibrate.

Many enone systems readily undergo 1,3-dipolar addition reactions. This applies to steroids, as illustrated by the formation of the fused pyrazole **16-4** from reaction of the enone **16-2** with diazomethane. Pyrolysis of that pyrazole results in extrusion of nitrogen to leave behind a methyl group at position 1 (**16-5**).

Reaction of the same starting material **15-2** with ethyl formate and sodium ethoxide, as in the case of the desalkyl series, proceeds to form the 2-formyl derivative **17-1** (Scheme 5.17). Exhaustive catalytic hydrogenation then gives the 2α-methyl derivative dromostanolone (**17-2**). The initial product, as in the case of the desoxy series, consists of the 2β-methyl derivative from approach of hydrogen from the open backside. Treatment with base epimerizes this to the favored 2α-methyl epimer **17-3** (dromostanolone), in which the newly introduced methyl group is equatorial.

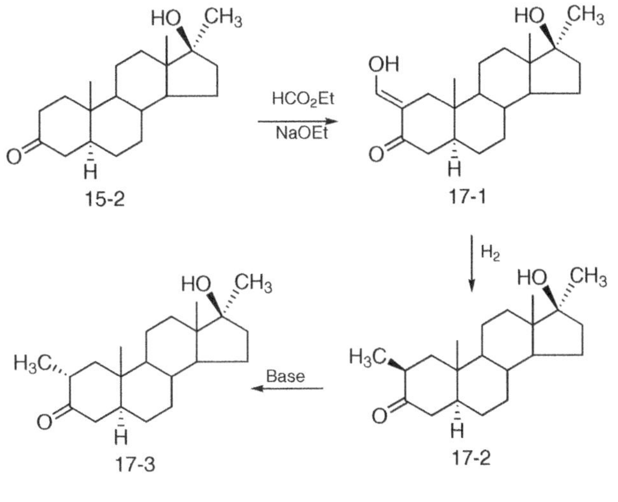

Scheme 5.17

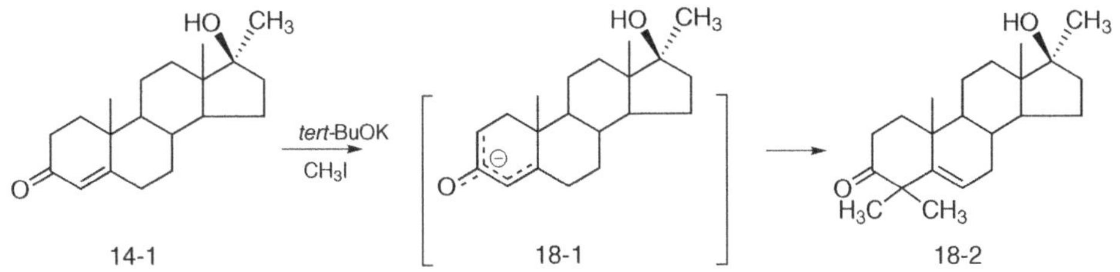

Scheme 5.18

Examples of reactions that involve carbanion-mediated alkylations have thus far all involved the formation of an enolate that led to reaction at C_2. Alkylation of 17-methyltestosterone, however, favors reaction at C_4. Treatment of that enone, **14-1**, with methyl iodide in the presence of the strong base potassium *tert*-butoxide leads to the formation of the 4,4-dimethyl derivative **18-2** (Scheme 5.18).

Position 4 in androstanes will undergo some reactions that are more typically associated with the chemistry of aromatic compounds. Treatment of methyltestosterone (**14-1**) with formaldehyde and thiophenol thus leads to the formation of the thiomethylated derivative **19-1** (Scheme 5.19). Raney nickel then desulfurizes the product, expelling the benzene ring in the process to afford 4,17-dimethyltestosterone (**19-2**).

Extension of the chain of double bonds in androstenes provides yet another site for modifying the basic structure. Bromination of the intermediate **14-2** that features the 1,4-dien-3-one system proceed on the allylic position at C_6 (**20-1**) (Scheme 5.20). Dehydrobromination by means of lithium carbonate in DMF leads to formation of the 1,4,6-trien-3-one system (**20-2**). Treatment of that intermediate with thioacetic acid proceeds to add one thiol group to each end of the conjugated systems to afford the bisthiolated product thiomestrone (**20-3**).

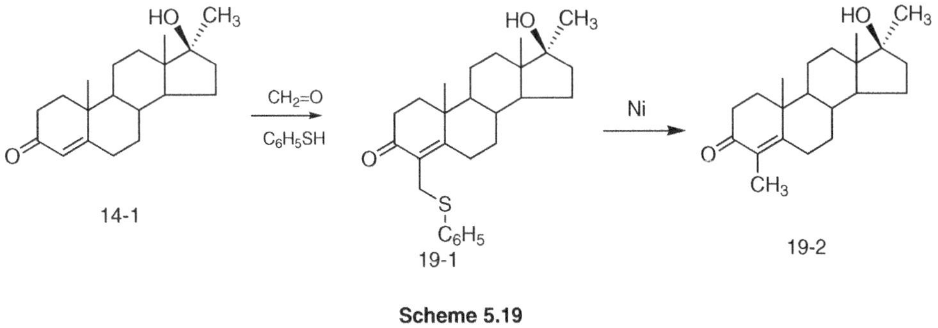

Scheme 5.19

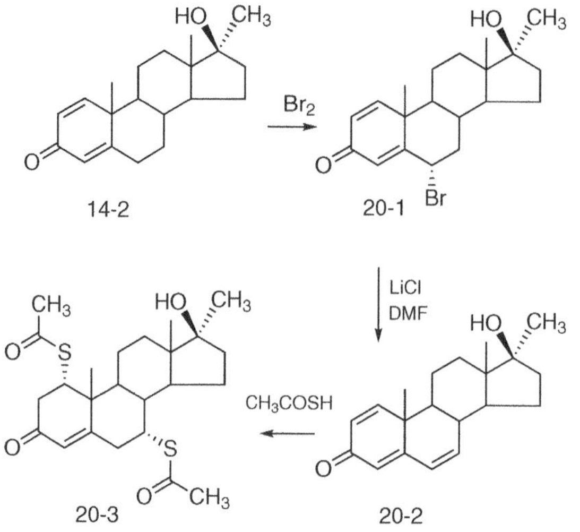

5.3.2.1.2 Fused Heterocycles

The β-dicarbonyl system in the formyl derivative of 17-methyltestosterone shows much the same reactivity as the same array in simpler compounds. Thus, reaction of the formyl derivative **17-1** with hydrazine fuses a pyrazole ring on to the steroid at positions 2,3 (**21-1**) (Scheme 5.21). This compound, stanazol, is one of the more frequently abused anabolic agents. Reaction of the same formyl derivative with hydroxylamine goes on to add an isoxazole ring to give danazol (**21-2**).

The reactions for forming fused heterocycles is quite general and applicable even to the 4,4-dimethyl analogue. Formylation of the 4,4-dimethyl starting material **18-2** proceeds to afford the 2-formyl derivative **22-1** (Scheme 5.22); the alternative, alkylation at C_4, is precluded by the two methyl groups at that site. Treatment with hydroxylamine interestingly goes on to form the isoxazole ring in spite of the increased crowding caused by the extra bulk at C_4 (**22-2**).

Construction of the simplest fused heterocyclic system that shows some anabolic activity starts by epoxidation of the olefin at C_2 in **23-1** with peracetic acid buffered with sodium acetate (**23-2**) (Scheme 5.23). Reaction of this intermediate with potassium thiocyanate and phosphoric acid, essentially thiocyanic acid, leads to diaxial opening of the oxirane to form the 3α-hydroxy-2β-cyanate **23-3**. Treatment of the product with base leads to formation of the β-thioepoxide **23-5**. Formation of that product can be rationalized by assuming that the first step involves removal by base of a proton on the hydroxyl group. That alkoxide then proceeds to add to the cyano group at C_2 to form a transient five-membered ring as in **23-4**. Sulfur then displaces the neighboring oxygen, closing the thioepoxide ring (**23-5**) with concomitant expulsion of a cyanate anion. The same sequence starting with the epoxide with reversed configuration of the oxirane leads to the analogue of **23-5** in which the thioepoxide ring bears the α-configuration.

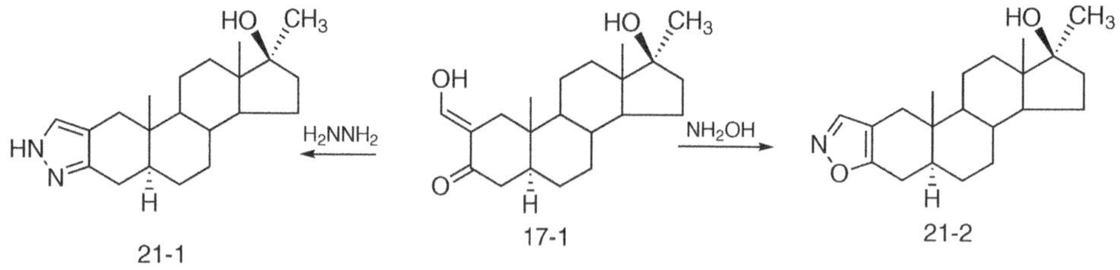

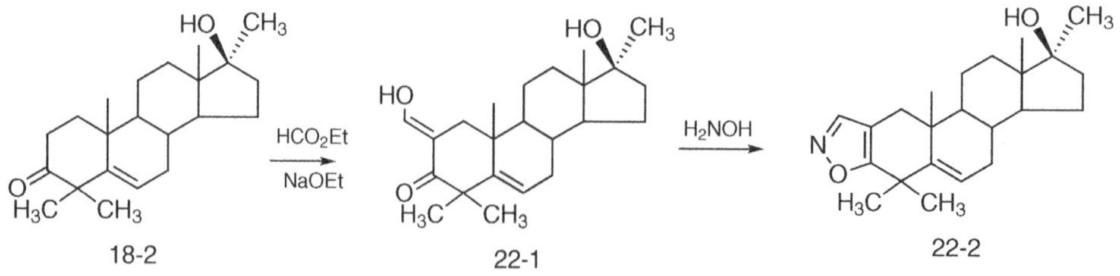

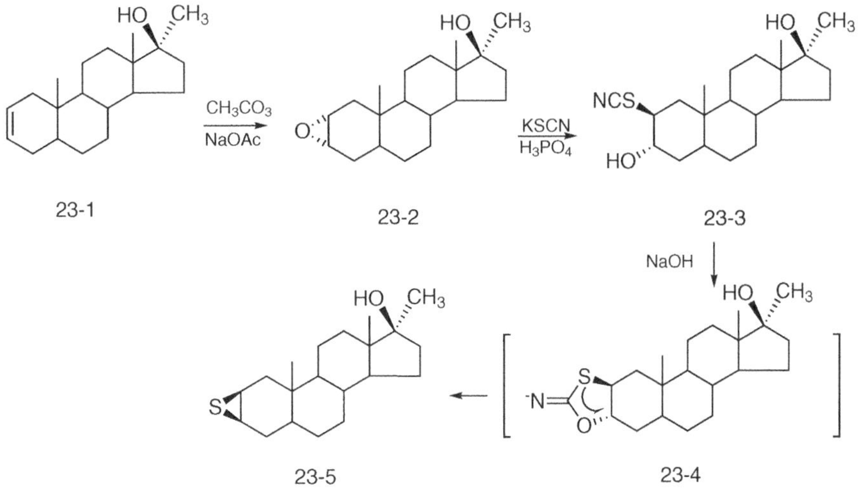

Scheme 5.23

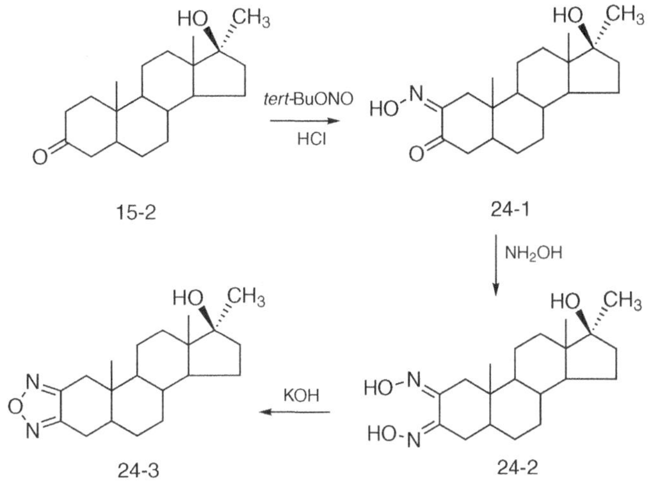

Scheme 5.24

Fusing a furazan on to ring A affords the potent anabolic androgen furazabol, yet another compound on the DEA List of Controlled Substances. This drug can be readily prepared by a three-step sequence. Reaction of androsta-17α-methylandrostan-17β-3-one (**15-2**) with *tert*-butyl nitrite in the presence of mineral acid in effect oxidizes the position next to the carbonyl, converting the methylene to a ketone masked as its oxime, **24-1** (Scheme 5.24). The ketone at C_3 is then itself converted to its oxime, **24-2**, by means of hydroxylamine. Heating that product with base closes the heterocyclic ring (**24-3**).

Fused heterocyclic rings can also be used to introduce acyclic substituents on ring A. The scheme for one such compound starts with the construction of an isoxazole fused on to ring A by the sequence depicted in Scheme 5.22. Thus, reaction of the 4,17-dimethylandrost-4-ene **19-2** with ethyl formate and sodium ethoxide leads to the 2-formyl derivative **25-1** (Scheme 5.25). Treatment of the product with hydroxylamine, again as in Scheme 5.22, gives the isoxazolo derivative **25-2**. Oxidation with *m*-chloroperbenzoic acid (*m*CPBA) leads to the formation of the α-4,5-oxirane **25-3**. Removal of the sole hydrogen on the heterocyclic ring with sodium methoxide leads to collapse of the heterocyclic ring so as to form the 2-cyano-3-keto derivative **25-4**. This β-cyano ketone then tautomerizes to its more stable enol form **25-5**.

5.3.3 Modifications on Ring B

5.3.3.1 C_6

The fairly standard scheme that is used to introduce a substituent at C_6 is also used in the pregnane and corticoid series. Application of the scheme to an androstane, specifically to introduce a methyl group, starts with the introduction of an

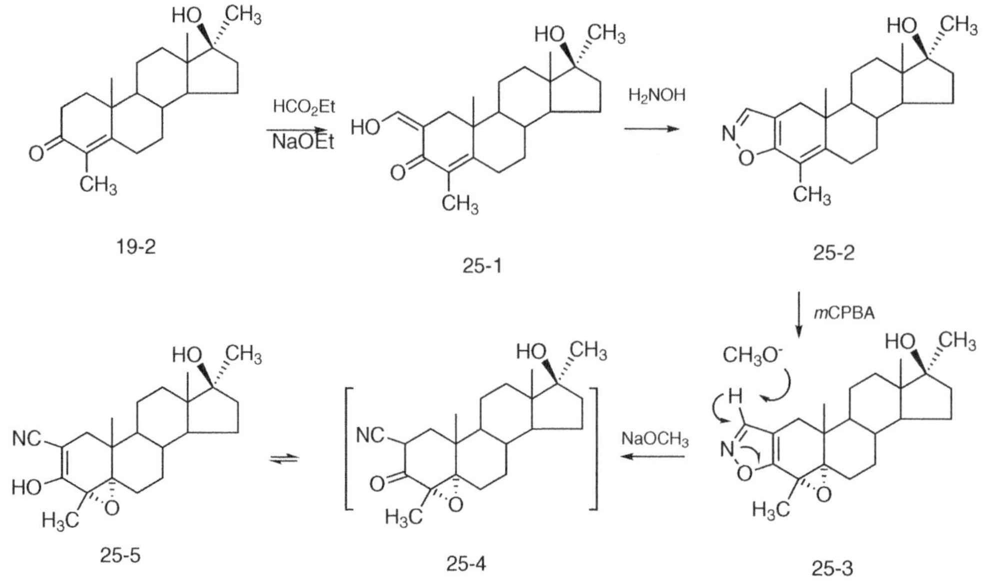

Scheme 5.25

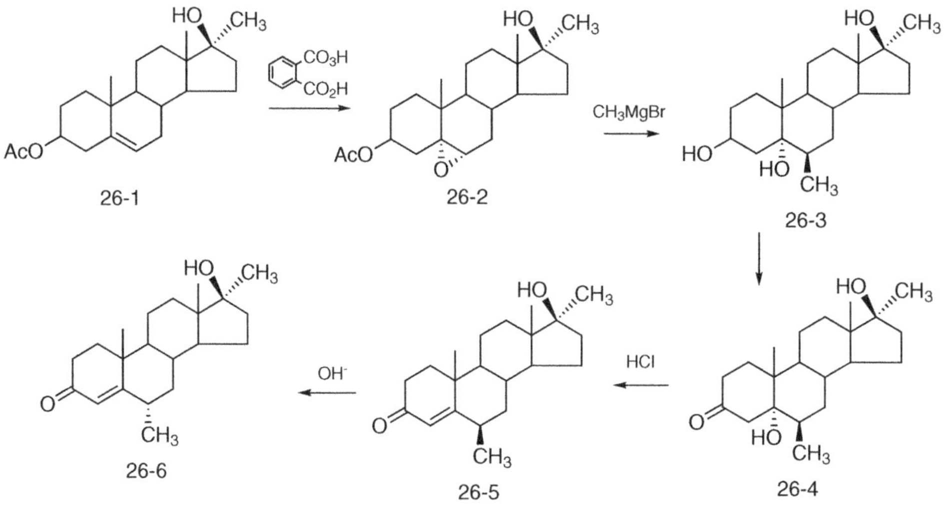

Scheme 5.26

oxirane in ring C with a peracid such as perphthalic acid (Scheme 5.26). The reagent attacks the olefin from the more open backside, forming the α-epoxide (**26-2**). Reaction of this intermediate with methylmagnesium bromide then opens the epoxide.

In the example at hand, the product of diaxial opening of the oxirane comprises the 5α-hydroxy-6β-methyl isomer **26-3**. Oxidation of the hydroxyl group at C_3 by the Oppenauer reaction affords the hydroxy ketone **26-4**. This loses water on heating with strong acid to form the 4-en-3-one **26-5**, in which the newly introduced methyl is configured in the axial β-position. The previously isolated methyl group is now located on an epimerizable position by virtue of its vinylogous relation with the ketone at C_3. Treatment with mild base thus causes the methyl group to isomerize to the preferred equatorial α-isomer, **26-6**.

The Vilsmeyer reaction is another reaction usually considered as a method for adding a substituent to aromatic rings. This transformation is, however, also applicable to cyclic enol ethers and more specifically the enol ether of a steroidal 4-en-3-one. The requisite enol ether, **27-1**, can be obtained by reaction of the 3,17-dione **1-6** with methanol in the presence of p-toluenesulfonic acid (TSA) (Scheme 5.27). Heating the resulting enol ether **27-1** with the methyl acetal of formaldehyde and phosphorus oxychloride in chloroform affords the 6-exo-methylene derivative **27-2** after workup. This product, named exemestane, is a potent inhibitor of the aromatase enzyme that catalyzes conversion of androstanes into ring-A-aromatic estrogens. This compound has consequently been used for treating estrogen-dependent cancers.

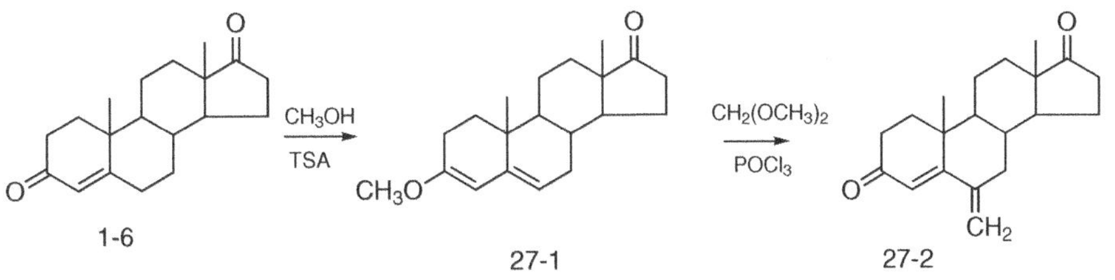

Scheme 5.27

5.3.3.2 C$_7$

The route used to introduce a substituent at C$_7$ in gonanes is also used in other steroid series. The key reaction in this sequence involves extending the enone system found in testosterone by one double bond. Treatment of 17-methyltestosterone (**14-1**) with chloranil removes hydrogen from ring C to afford the 4,6-diene **28-1** (Scheme 5.28). Reaction of that dienone with methylmagnesium bromide in the presence of a cuprous salt soluble in an organic solvent affords a mixture of the two isomeric products from conjugate addition (**28-2**). Separation of the crude product yields the predominant epimer bolasterone (**28-3**) and the lesser epimer calusterone (**28-4**). The ring position to which the new methyl groups are attached, C$_7$, is fairly remote from the carbonyl group and therefore not subject to epimerization. The axial epimer, blasterone is one of the most potent known anabolic–androgenic agents. The compound with the equatorial 7-methyl group, calusterone, has been used as an antineoplastic agent.

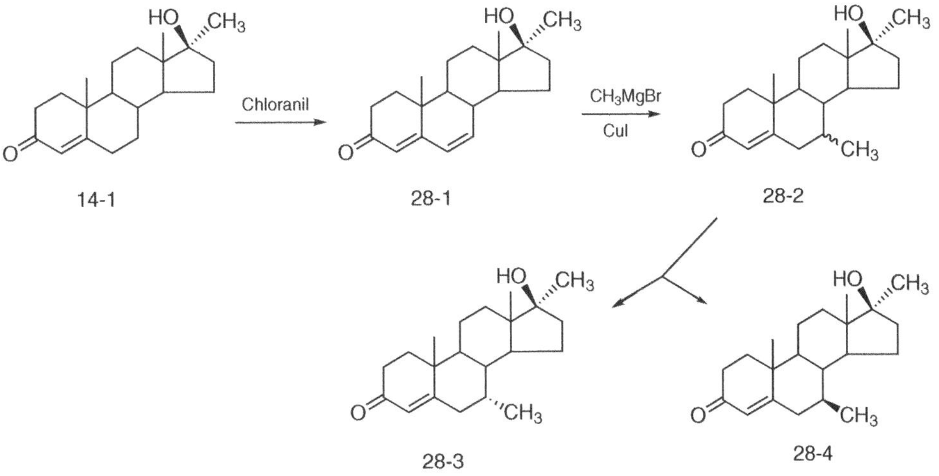

Scheme 5.28

5.3.4 Modifications on Ring C

The starting material for steroids modified in ring C depends on the presence in the molecule of a ketone or hydroxyl group at C$_{11}$. Microbiological introduction of a hydroxyl group at that carbon in progesterone, discussed in more detail in Chapter 7, provided the key for production of corticosteroids. Scission of the side chain in cortisone (**29-1**) by sodium bismuthate provides one source for adrenosterone (**29-2**) (Scheme 5.29). More recent work has identified organisms, such as *Aspergillus tamarii*, that add a hydroxyl group at C$_{11}$ when fermented with testosterone (**29-3**). Oxidation of that fermentation product with chromium trioxide also affords adrenosterone.

Adding a fluorine atom at C$_9$ significantly enhances the potency of corticosteroids (see Chapter 7). The scheme used to introduce that function in an androstane (**30-6** → **31-4**) closely mirrors that which is used for the structurally more complex compounds.

The sequence of reactions nicely points out the relative reactivity of carbonyl groups at positions 3, 11 and 17. Reaction of the triketone **29-2** with a controlled amount of pyrrolidine leads to the formation of the enamine from the most reactive ketone, that at C$_3$ (**30-1**) (Scheme 5.30). Treatment of this intermediate with methylmagnesium bromide leads to exclusive addition to C$_{17}$. Although the ketone at C$_{11}$ is virtually inert to addition reactions, it is subject to reduction. Reaction of **30-2** with lithium aluminum hydride thus leads to the corresponding β-hydroxy derivative. The enamine is then removed by acid hydrolysis (**30-4**). Reaction of the newly formed alcohol with *p*-toluenesulfonyl

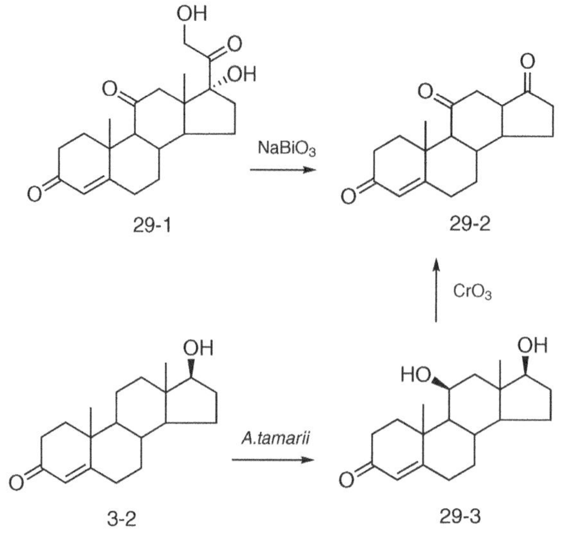

Scheme 5.29

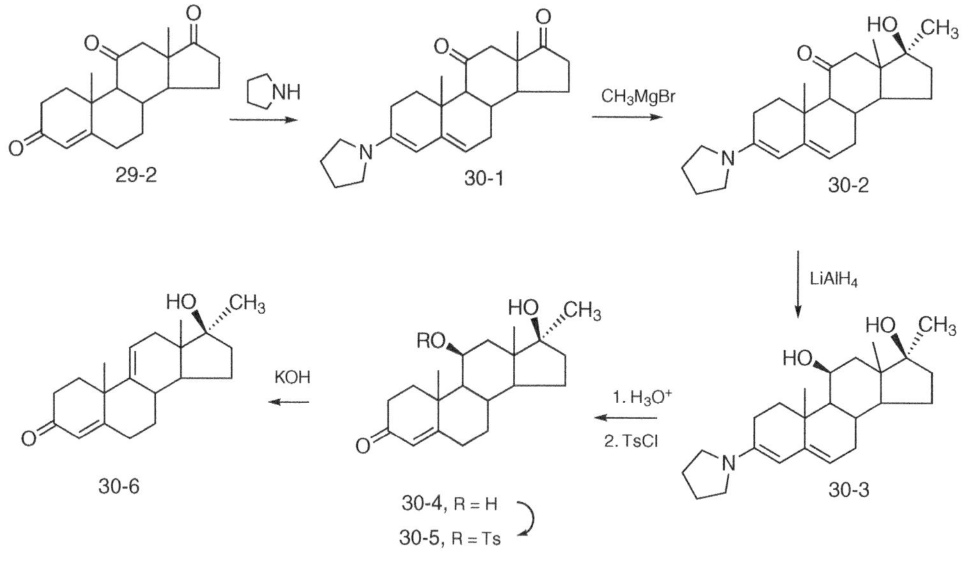

Scheme 5.30

chloride converts the hydroxyl to the leaving group toluenesulfonate (**30-5**). Treatment of intermediate **30-5** with base leads to elimination of toluenesulfonic acid and formation of the 9(11)-olefin.

The next step in the sequence calls for the formation of a 9(11)β-epoxide; direct oxidation with a peracid would probably give the α-epoxide from approach of reagent from the less hindered side. The first reaction in three-step procedure that leads to addition of fluorine at C_9 starts with treatment of the olefin **30-6** with aqueous N-bromoacetamide, in essence hypobromous acid. Formation of the product **31-2** can be rationalized by assuming that the initial step comprises formation of an α-bromonium adduct such as **31-1** (Scheme 5.31). Hydroxide ion then opens the ring to give the observed product **31-2** with bromine and hydroxide each disposed in an axial configuration. Reaction of the product with base involves displacement of halogen by the base-formed alkoxide on the adjacent carbon atom to give the β-oxirane **31-3**. The diaxial epoxide opening rule dictates both the regio- and stereochemistry of the next step. Treatment of the oxirane with anhydrous hydrogen fluoride in THF opens the epoxide to afford the 9α-fluoro-11β-hydroxyandrostane **31-4**, also known as the androgenic–anabolic agent fluoxymestrone. Attempts to open the ring with hydrogen fluoride under other conditions lead to inseparable mixtures, pointing to the probable intervention of an HF–THF complex.

5.3.5 Modifications on Ring D

An androstene almost devoid of functionality forms a major part of boar pheromone and has been used in breeding swine. (The author observed many years ago that androstanes with very few polar groups give forth a strong odor that brings to mind long-neglected urinals!)

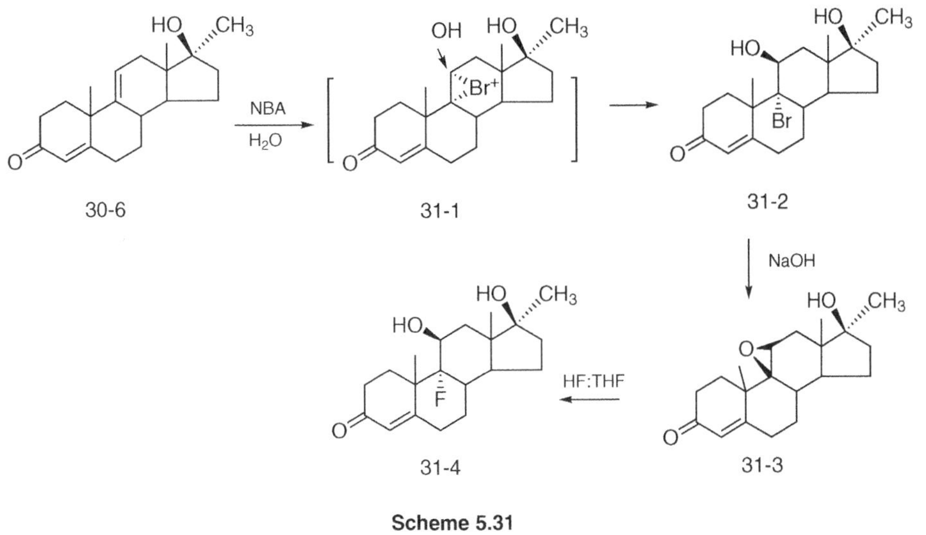

Scheme 5.31

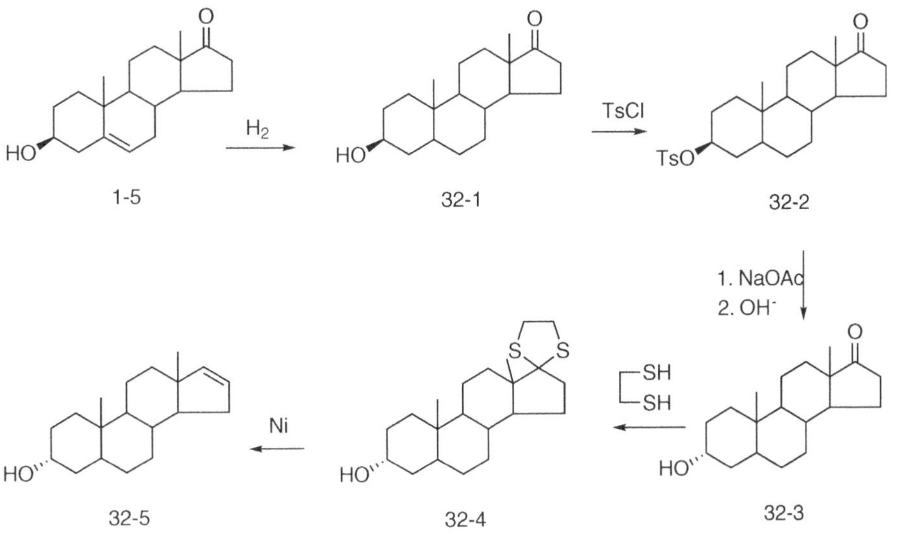

Scheme 5.32

The synthesis of this agent starts with catalytic hydrogenation of DHEA (**1-5**) to give the fully reduced androstane **32-1** (Scheme 5.32). The hydroxyl group at C_3 is next converted to a good leaving group by reaction with *p*-toluenesulfonyl chloride (**32-2**). Treatment of that intermediate with sodium acetate then displaces the toluenesulfonate with concomitant inversion of C_3. Saponification yields the corresponding alcohol **32-3**. Treatment of this intermediate with ethylenedithiol gives the thioacetal **32-4**. Reaction of the thioacetal with moderately active Raney nickel causes abstraction of sulfur to leave an olefin in ring D (**32-5**).

Knoevenagel condensation provides the key to adding an additional substituent in ring D. Reaction of DHEA acetate (**33-1**) with acetaldehyde and sodium ethoxide leads to the 16-ethylidine product **33-2**. The stereochemistry of the new olefin is irrelevant, as it will be erased in the next step. Catalytic reduction proceeds by addition of hydrogen from the more open backside of the molecule to give the 16β-ethyl derivative **33-3**. This quasi-axial position of the product actually represents the energetically disfavored configuration. Treatment of the product with lithium aluminum hydride again leads to addition of hydrogen to the ketone from the backside to give the 17β-alcohol; the acetate at C_3 is also reduced to a hydroxyl group to afford **33-4**.

The synthesis of a topical anti-inflammatory, anti-allergic agent that is inactivated in the bloodstream starts with the scission of the dihydroxyacetone side chain of the corticosteroid fluoroprednisolone (**34-1**) (see Chapter 7, Scheme 7.17) with sodium bismuthate (Scheme 5.34). Reaction of the product, **34-2**, with methanethiol affords the thioacetal **34-3**. The acetal extrudes one of the thiomethyl groups under acidic conditions to form the enol thioether **34-4**. Treatment of that product with ethanethiol restores the thioacetal (**34-5**). The stereochemistry at C_{17} is defined by addition of ethanethiol from the more open backside of the thioenol **34-4**.

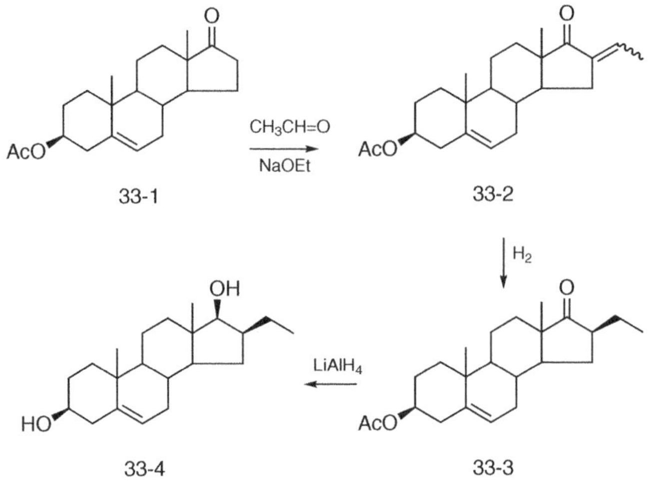

Scheme 5.33

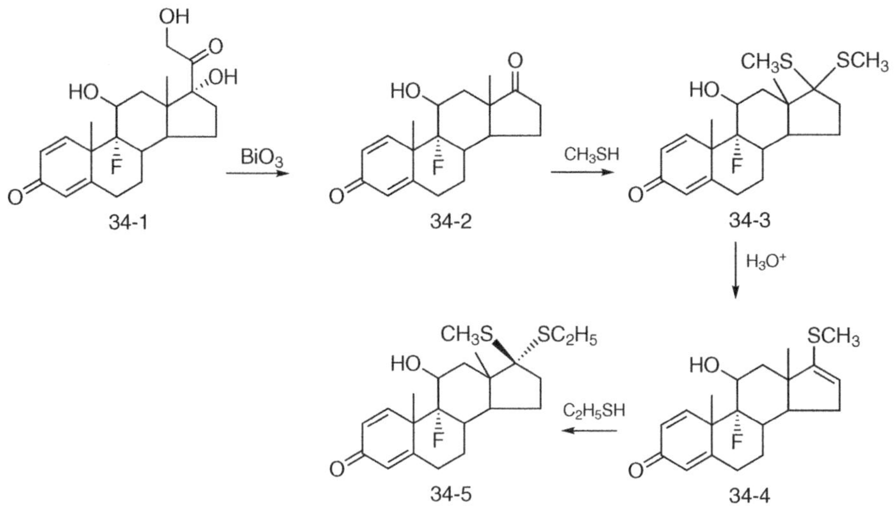

Scheme 5.34

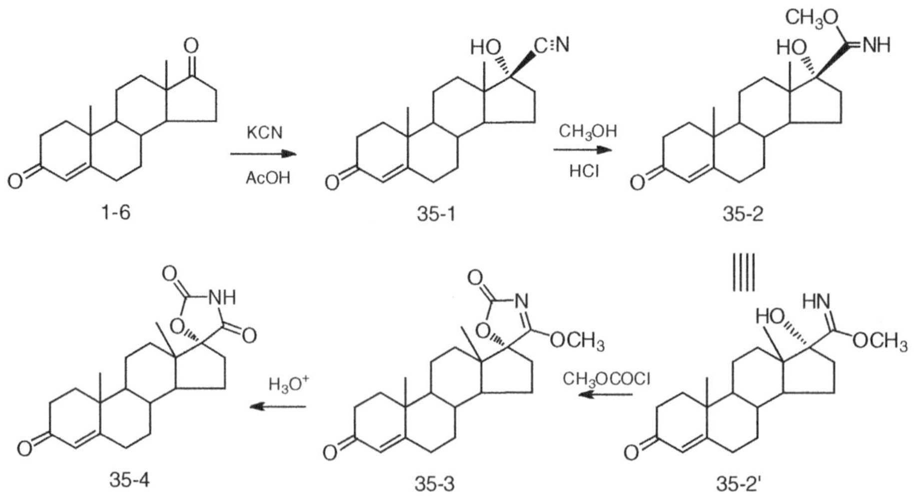

Scheme 5.35

Addition of cyanide to androst-4-ene-3,17-dione gives the cyanohydrin **35-1** with a 17β-nitrile (Scheme 5.35). The stereochemical result is a reflection of the reversible nature of the transformation in which the sterically favored isomer predominates. Reaction of the cyanohydrin with methanolic hydrogen chloride converts the nitrile to the imino ether **35-2** (also shown as its rotamer **35-2′**). Methyl chloroformate then closes the ring to form the spirooxazoline ring (**35-3**). Treatment of this intermediate with acid hydrolyzes the imino ether function to give the corresponding oxazolidine-2,4-dione **35-4**.

5.4 17-Spirobutyrolactone Aldosterone Antagonists

Aldosterone, essentially a corticosteroid in which the methyl group at C_{18} is oxidized to a carboxaldehyde, was the last endogenous steroid to be isolated in a form suitable for structural studies. This hormone controls via its action on the kidney both body electrolyte balance and blood volume. A number of modified androstanes, all of which feature a spirobutyrolactone at C_{17}, act as aldosterone antagonists Those 17-spirobutyrolactones consequently show diuretic and concomitant antihypertensive activity.

The synthesis of the simplest agent in the group first involves treatment of the ethisterone intermediate **13-3** with a large excess of methylmagnesium bromide in order to effect metal interchange between the Grignard reagent and the acidic terminal proton on the acetylide side chain. Addition of carbon dioxide to the reaction mixture leads to the formation of the corresponding propargylic acid **36-2** (Scheme 5.36). Catalytic hydrogenation then reduces the side chain to a propionic acid. Lactonization of the hydroxyl acid, for example with acetic anhydride, forms the spirolactone **36-4**; the hydroxyl group at C_3 is acetylated in the process (**36-4**). Hydrolysis of the acetate followed by Oppenauer oxidation then affords the prototype aldosterone antagonist spirolactone (**36-6**).

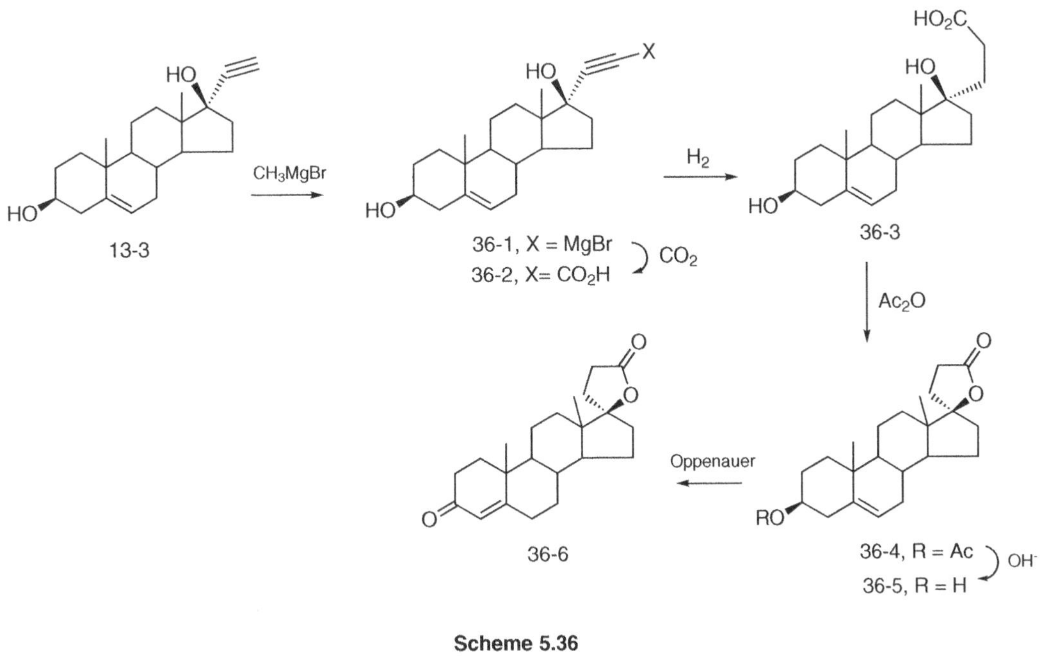

Scheme 5.36

The initial member of the group, **36-6**, is active only by injection. A relatively small change to the structure leads to an agent that is orally active. As is the case with other androstanes, treatment of that compound with chloranil abstracts hydrogen from ring B, extending the conjugation by one more olefinic bond (**37-1**) (Scheme 5.37). Reaction of that intermediate with thiolacetic acid leads to addition of this nucleophile to the end of conjugated system, C_7, to give the orally active drug spironolactone (**37-2**). It is notable that the active species actually consists of the ring-opened hydroxyl acid.

The ylide from trimethylsulfonium iodide is best known for converting carbonyl groups to homologous epoxides. When allowed to react with a conjugated ketone, this reagent converts a terminal olefin into a fused cyclopropane. The reaction with the intermediate **37-1** can be visualized by assuming that the initial step with involves addition of the ylide from the trimethylsulfonium iodide and sodium hydride to the end of the enone system with concomitant movement of the negative charge on to carbonyl oxygen (**37-3**). Return of the charge leads to attack of the charge on the carbon of the ylide, resulting in the formation of the cyclopropyl group and expulsion of neutral dimethyl sulfide. The product of this reaction, canrenone (**37-4**), is sold as its ring-opened potassium salt.

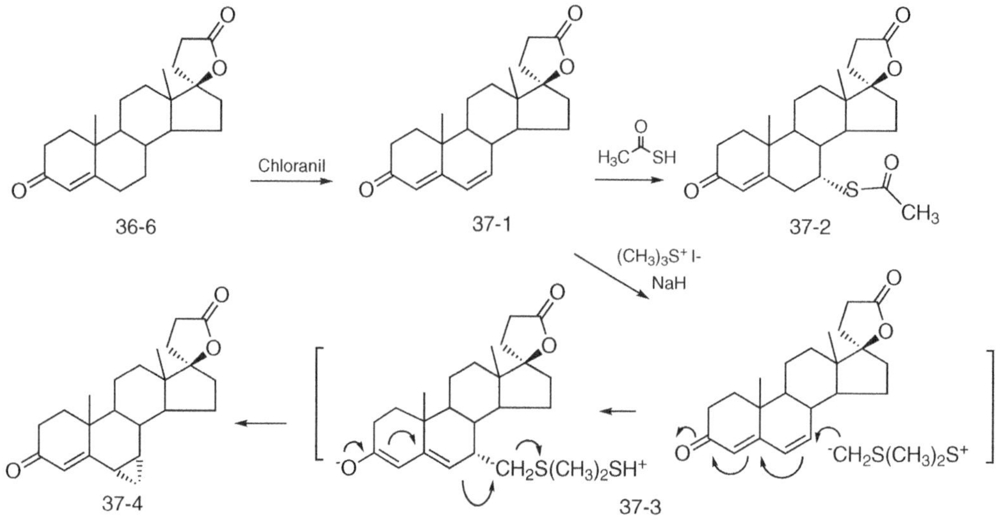

84 *Steroid Chemistry at a Glance*

Scheme 5.37

The overall sequence for forming the analogue that bears a carbethoxy group at C_7 would seem to be relatively straightforward as it involves adding the carbon atom by way of cyanide followed by methanolysis of the product. However, the actual course of the process is anything but straightforward and illustrates the effect on reactions of the constrained nature of the steroid nucleus. The first step comprises addition of cyanide to the dienone system in **37-1**. The reagent apparently adds first at C_7 at the end of the conjugated system. The remaining 3-on-4-ene is sufficient electrophilic to accept a second cyanide to form an intermediate such as **38-1** (Scheme 5.38).

This intermediate (**38-1**) is not observed as it undergoes internal addition of the charge at C_4 to the nitrile at C_7. The observed product from this last reaction, **38-2**, is isolated and then subjected to acid hydrolysis to the corresponding ketone **38-3**. Reaction of **38-3** with sodium methoxide leads initially to addition of the bridging ketone **38-4**. The charge on oxygen on C_3 then returns, expelling the remaining nitrile group, forming mexrenone (**38-6**).

A similar set of reactions starting with the spirobutyrophenone **39-1**, which incorporates in addition a double bond at positions 9(11) in ring C, proceed in a perfectly straightforward manner (Scheme 5.39). This may be due in part to some subtle change in the shape of the molecule imparted by the extra double bond. Reaction with cyanide proceeds to add a nitrile at the end of the conjugated system (**39-2**). Reduction of the newly added group with diisobutylaluminum converts the nitrile to the corresponding imine. Acid hydrolysis then converts that function to the carboxaldehyde **39-3**. That functional group is next oxidized to a carboxylic acid; treatment of the oxidation product with diazomethane

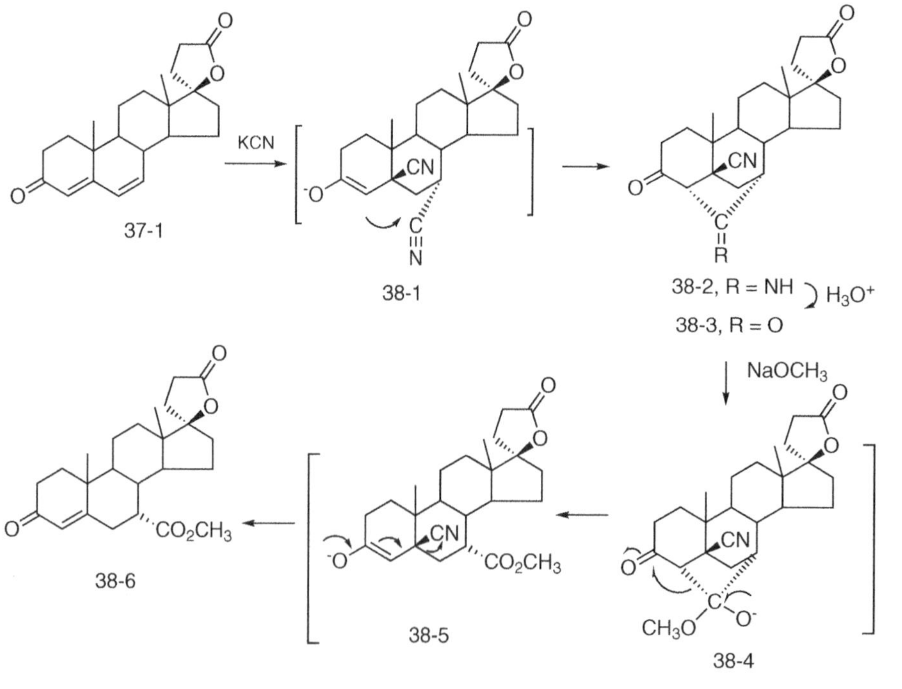

Scheme 5.38

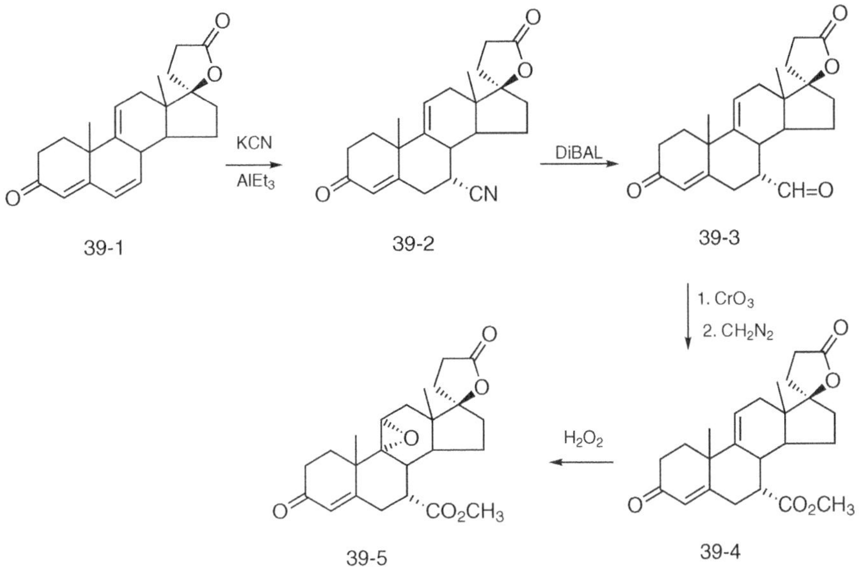

Scheme 5.39

affords the methyl ester, **39-4**. Oxidation of the olefin at $C_{9(11)}$ with the higher electron density with hydrogen peroxide gives the 9(11)-epoxide eplerenone (**39-5**).

5.5 Some Drugs Based on Androstanes

5.5.1 Androgens

The androstanes described in Schemes 5.3–Scheme 5.32 generally act as anabolic–androgenic agents. Prescriptions for treating cases of androgen deficiency comprise a minor but important use for those agents. The major legitimate market for drugs approved by the FDA relates to the compounds' anabolic, nitrogen-sparing activity. These agents are therefore often used for treating conditions associated with severe tissue loss, such as severe burns. The large illicit market from athletes and body-builders has already been noted. Virtually all these drugs are now no longer protected by patents and are as a result manufactured by several firms. The majority of the drugs are as a result available under more than one trade name. The generic (USAN) names for some approved drugs include (in alphabetical order) calusterone (**28-4**), danazole (**21-2**), fluoxymestrone (**31-4**), methyltestosterone (**13-2**), oxymetholone (**17-1**) and stanozole (**21-1**). Some non-classical applications include the use of danazole to treat endometriosis as a result of its anti-gonadotropic properties and, for the same reason, calusterone in clinical trials against breast cancer.

5.5.2 Spirobutyrolactones

Thiazide diuretics, typified by hydrochlorothiazide, tend to cause excess urinary excretion of potassium over sodium ion, upsetting the normal serum concentration of these ions. Administration of spirobutyrolactones, on the other hand, leads to a more physiological balance of those ions in urine. As a result, these drugs are often termed potassium-sparing diuretics. The pioneer orally active spirobutyrolactone (**37-2**) is still on the market as a diuretic under the generic name spironolactone and the trade name Aldactone. Now that the patent has expired, the agent is now sold as a generic drug by many manufacturers, and as a result is available under a plethora of trade names.

6
Pregnanes, Part 1: Progestins

6.1 Biological Activity

The role of estradiol in the female reproductive cycle is outlined briefly in Chapter 3. Progesterone comprises the second of the steroidal hormones involved in this cycle. Serum concentrations of this steroid, like those of its partner, demonstrate an analogous increase and ebb over the course of the menstrual cycle. To recapitulate levels remain relatively low until ovulation occurs at mid-cycle. At that point, the follicle, a blister-like structure on the ovary from which the ovum was released, undergoes a morphological change. The new structure, now termed a corpus luteum, is now the principal source of progesterone. That transformation then leads to rising levels of progesterone, which in turn cause proliferation of the lining of the uterus intended to provide a setting for a fertilized ovum. In the case where the ovum is sterile, the lining collapses and is shed along with menstrual blood; the concomitant ablation of the corpus luteum causes progesterone levels to decrease. The scenario changes when a fertilized ovum implants in the uterine lining: the corpus luteum now persists and secretes increasing levels of progesterone for the full term of the pregnancy. Those elevated levels of the hormone block further ovulation. Depending on the species, levels can stay high to the end of lactation.

It is of passing interest that the ovulation-inhibiting role of progesterone prompted the research that led to the oral contraceptives. The analogous activity of estrogens was not appreciated at the time. Some modified derivatives of progesterone, more potent than the endogenous hormone, had at one time been used in oral contraceptives although they have now been largely displaced by the 19-norprogestins. Other clinical uses for progesterone and its derivatives devolve largely on treating conditions caused by deficient levels of the hormone. A serious drawback to the use of progesterone proper is the fact that it must be administered by injection since it lacks oral activity. The search for orally active analogues has served as an added stimulus for the synthesis of progesterone analogues.

6.2 Sources of Progesterone

6.2.1 From Phytochemicals

The process for producing progesterone from soybean-derived stigmasterol is outlined in Chapter 2.

Pregnenolone from diosgenin can be converted to progesterone in two straightforward steps. Thus, catalytic hydrogenation of **1-1** leads to the dihydro derivative **1-2** (Scheme 6.1). Oppenauer reaction provides the carbonyl group

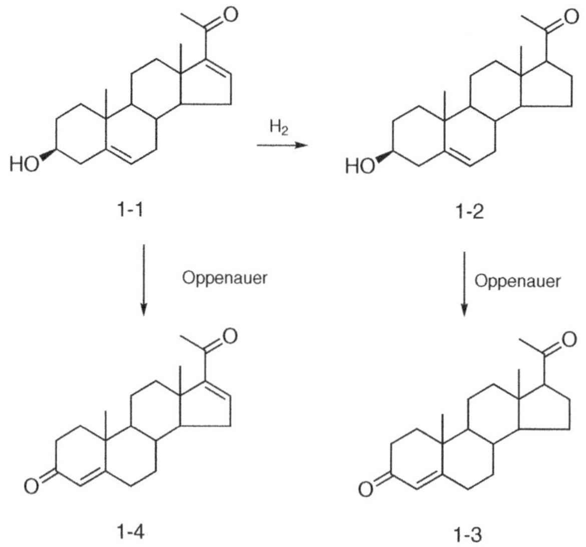

Scheme 6.1

Steroid Chemistry at a Glance   Daniel Lednicer
© 2011 John Wiley & Sons, Ltd

Pregnanes, Part 1: Progestins **87**

at C_3; the double bond then shifts into conjugation to provide progesterone (**1-3**). The syntheses of some of the progesterone derivatives with modifications in ring D require the presence of a double bond at C_{16}. In the same vein, 16-dehydroprogesterone (**1-4**) can be obtained from 16-dehydropregnenolone (**1-1**) by Oppenauer oxidation.

6.2.2 By Total Synthesis

Steroids have posed a challenge to synthetic organic chemists dating back almost to the day when their structure was fully elucidated. Total syntheses have been developed and presented in this book for each of the increasingly complex structural classes. The total synthesis of progesterone, from the laboratory of the late W. S. Johnson, is rightfully named biomimetic. The key step in this synthesis comprises a cascade of ring closures reminiscent of the corresponding step in the cyclization of squalene to form lanosterol (see Chapter 2, Scheme 2.3).

Construction of the polyene chain first involves condensation of the aldehyde **2-2** with the ylide from treatment of the phosphonium salt **2-1** with phenyllithium (Scheme 6.2). The *trans* configuration of the new double bond in the product **2-3** represents the normal course for the Wittig reaction. Treatment with acid then hydrolyzes the acetal groups to reveal the 1,4-dione function (**2-4**). This cyclizes to the cyclopentenone **2-5** on treatment with base. Cyclization in the opposite sense is apparently not problematic. Reaction with methyllithium then affords intermediate **2-6**, which now includes a fairly labile tertiary allylic methylcarbinol.

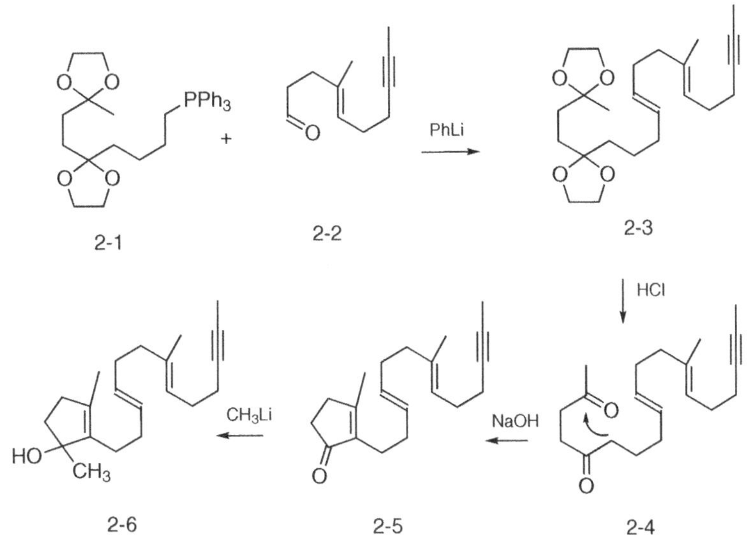

Scheme 6.2

The molecule is now set for the key biomimetic cascade. Thus, treatment of **2-5** with trifluoroacetic acid leads to ejection of the carbinol to leave behind a carbocation (**3-1**) (Scheme 6.3).

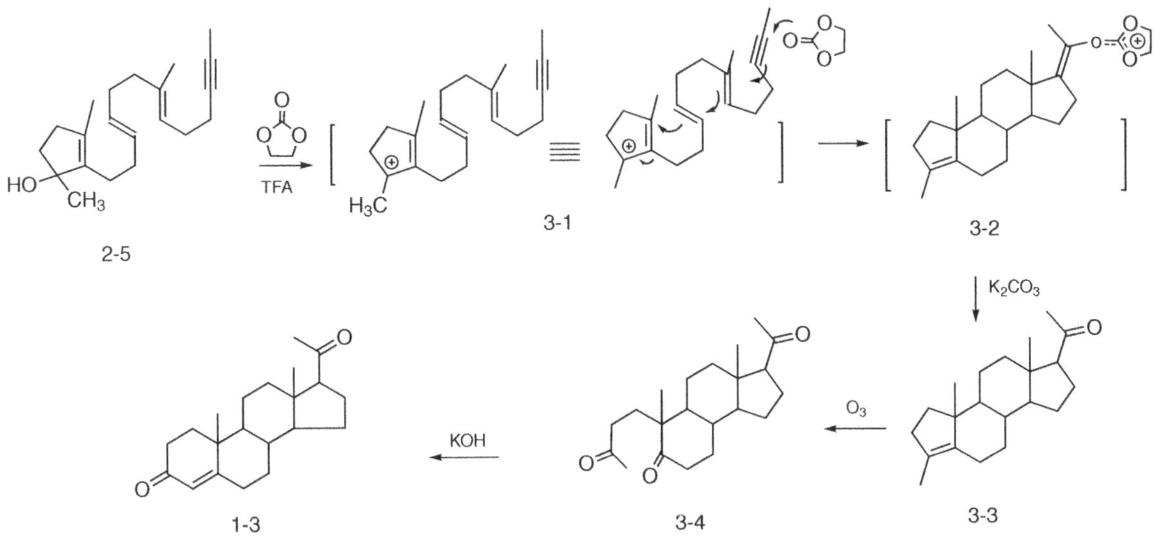

Scheme 6.3

The ethylene glycol acetal of carbonic acid present in the reaction medium serves a dual function: the carbonyl oxygen in this reagents provides the electron that neutralizes the carbocation at the end of the cascade (**3-2**) and the acetal, in addition, provides the oxygen that will form the C_{20} carbonyl group in progesterone. Neutralization of the reaction mixture with potassium carbonate yields the intermediate **3-3**, which includes all progesterone carbon atoms. The conformation of the polyene **2-6** is such as to cause the new rings to assume steroidal stereochemistry. It remains to expand ring A to a cyclohexanone. Ozonization of **3-3** followed by oxidative workup of the ozonide affords the diketone **3-4**. Treatment of that intermediate with base leads to aldol condensation that forms a cyclohexanone and thus racemic progesterone (**1-3**).

6.2.3 From Dehydroepiandrosterone (DHEA) Acetate

The preparation of dehydroepiandrostenolone from 16-dehydropregnenolone is described in the preceding chapter. It is also possible to run that intraconversion in reverse. Thus, reaction of DHEA acetate (**4-1**) with potassium cyanide in acetic acid leads to the cyanohydrin **4-2** (Scheme 6.4). Reaction of that product with a dehydrating reagent such as phosphorus oxychloride leads to the corresponding 16-dehydro derivative **4-3**. The requisite extra carbon is then incorporated by addition of methylmagnesium bromide to the nitrile. The initially formed imine is hydrolyzed to a ketone on workup; the acetate at C_3 is also hydrolyzed to a free hydroxyl in the process to yield 16-dehydropregnenolone (**1-1**).

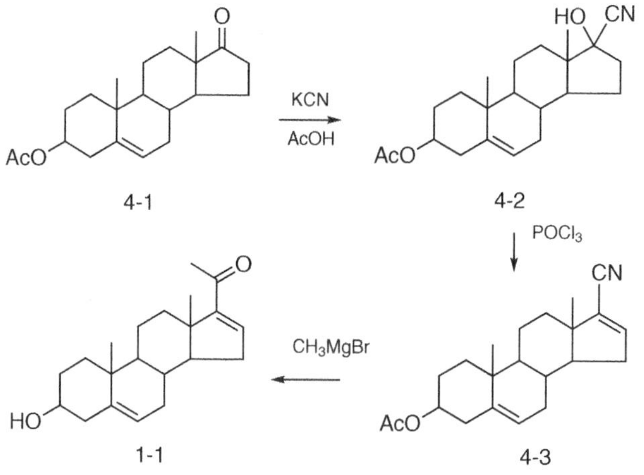

Scheme 6.4

6.3 Modified Pregnanes

6.3.1 17-Hydroxy and Acyloxy Derivatives

The derivative of progesterone that includes a hydroxyl group at C_{17} is a naturally occurring metabolite, esters of which are more potent than the parent molecule and are also orally active. A method for including this modification is discussed at this point since it occurs in several analogues. A variety of peracids can be used to convert isolated olefinic linkages to their epoxides. Double bonds conjugated with a carbonyl group are, however, more commonly epoxidized by means of alkaline hydrogen peroxide; isolated double bonds are inert to this reaction since it is initiated by conjugate addition of a hydroperoxy anion to the enone. Treatment of pregnenolone (**1-1**) under these conditions thus gives the product **5-1** (Scheme 6.5). The α-stereochemistry of the oxirane reflects attack of the reagent from the backside. Reaction of this intermediate with hydrogen bromide proceeds to the product from diaxial opening of the epoxide **5-2**. Catalytic hydrogenation in the presence of ammonium acetate as buffer removes the halogen to give the 17α-hydroxy derivative **5-3**. Forcing conditions are required to acylate the very sterically hindered carbinol at C_{17}. Treatment of this last intermediate with acetic anhydride catalyzed by *p*-toluenesulfonic acid thus gives the 17α-acetate **5-4**. Saponification proceeds selectively at C_3 to afford the alcohol **5-5**. Oppenauer reaction then oxidizes that hydroxyl and shifts the olefin into conjugation to give 17α-hydroxyprogesterone acetate (**5-6**).

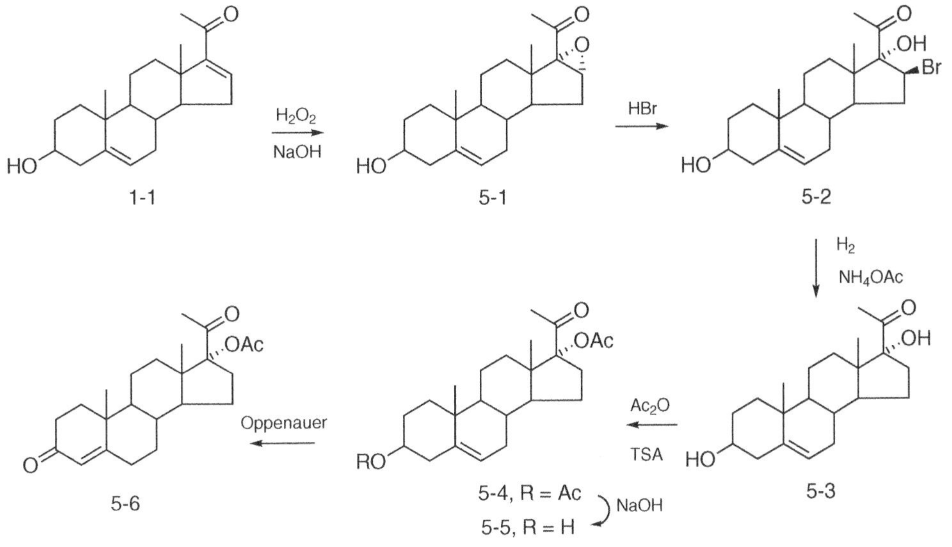

Scheme 6.5

6.3.2 Modifications on Ring A

The sterically relatively open carbonyl group at C_3 is significantly more reactive than its hindered counterpart at C_{20}. Reaction of 16-dehydroprogesterone (**1-4**) with benzenethiol takes a different course at the two conjugated ketones present in the compound. Reaction at C_3 proceeds to form an enol thioether; reaction with the unsaturation in ring D, on the other hand, consists of conjugate addition of the thiol at a distance from the hindered carbonyl (**6-1**) (Scheme 6.6).

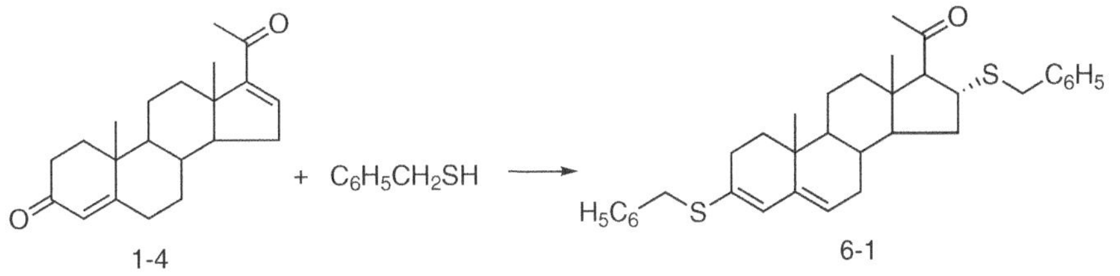

Scheme 6.6

This marked difference in reactivity is also observed when progesterone is reacted with pyrrolidine. This reaction proceeds to yield the enamine **7-1** from reaction at C_3 (Scheme 6.7). Reaction of that enamine with perchloryl fluoride ($FClO_3$) followed by hydrolysis of the enamine gives the 4-fluoro derivative **7-2** of undefined stereochemistry. Treatment with acid shifts the olefin into conjugation to afford 4-fluoroprogesterone (**7-3**).

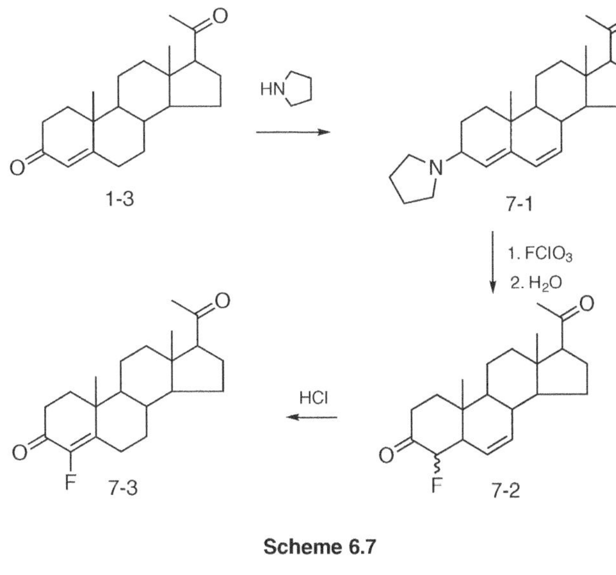

Scheme 6.7

6.3.3 Modifications on Ring B

The 17-ethynyl derivative of testosterone, ethisterone (see Chapter 5, Scheme 5.13), does not contain the 17-acetyl group that characterizes derivatives of progesterone; it can, however, be classed as a progestin by both its biological activity and systematic name. The sequence for adding a methyl group to this molecule is not only typical for that applicable to progesterone and its derivatives, it is also used to add a methyl group to androstanes (see Chapter 5, Scheme 5.26).

The sequence for preparing the 6β,21-dimethyl analogue of ethisterone starts with epoxidation of the unsaturation at C_5 with perphthalic acid to afford the epoxide **8-2** (Scheme 6.8). Reaction with methylmagnesium bromide inserts a methyl group at C_6; both the stereo- and regiochemistry of the product **8-3** follow from diaxial opening of the oxirane. The three hydroxyl groups in the product are then protected from conditions in the next step by conversion to tetrahydropyranyl acetals by reaction with dihydropyran (not depicted). Treatment of the resulting mixture of diastereoisomeric hydropyran acetals with sodium amide generates an anion at the terminus of the ethynyl side chain. That anion is then alkylated with methyl iodide. Workup with aqueous acid then hydrolyses the ethers to afford the triol **8-4**. Oxidation of this intermediate with a chromium trioxide–pyridine complex generates the carbonyl group at C_3. The resulting β-hydroxy ketone readily dehydrates when treated with strong acid. The methyl group in the first-formed product occupies the disfavored axial β-configuration. This substituent then epimerizes to the more stable α-isomer under the strongly acidic conditions.

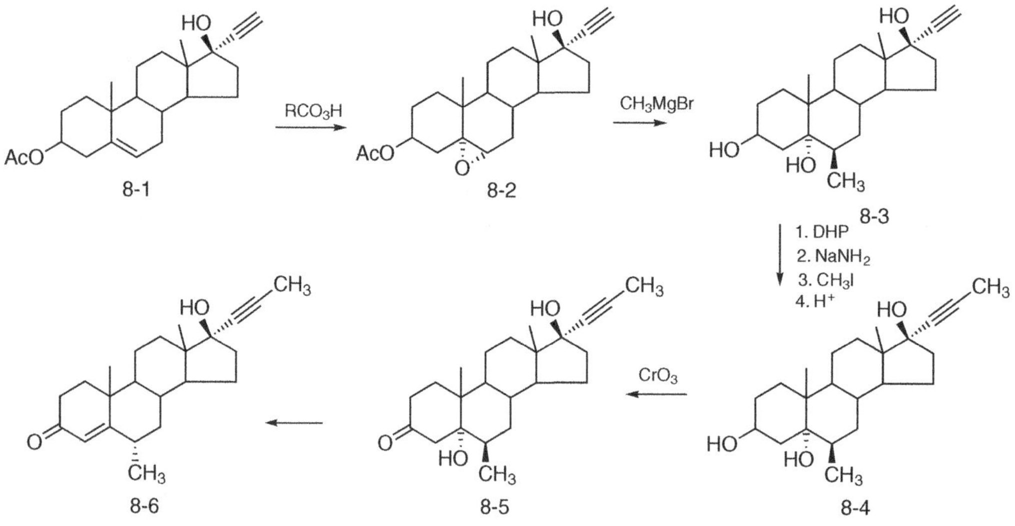

Scheme 6.8

An analogous sequence starting with 17α-hydroxyprogesterone (**9-1**) leads to the corresponding 6-methyl analogue (Scheme 6.9). The first step comprises protection of the carbonyl group at C_3 and C_{20} as their ethylene acetals. That

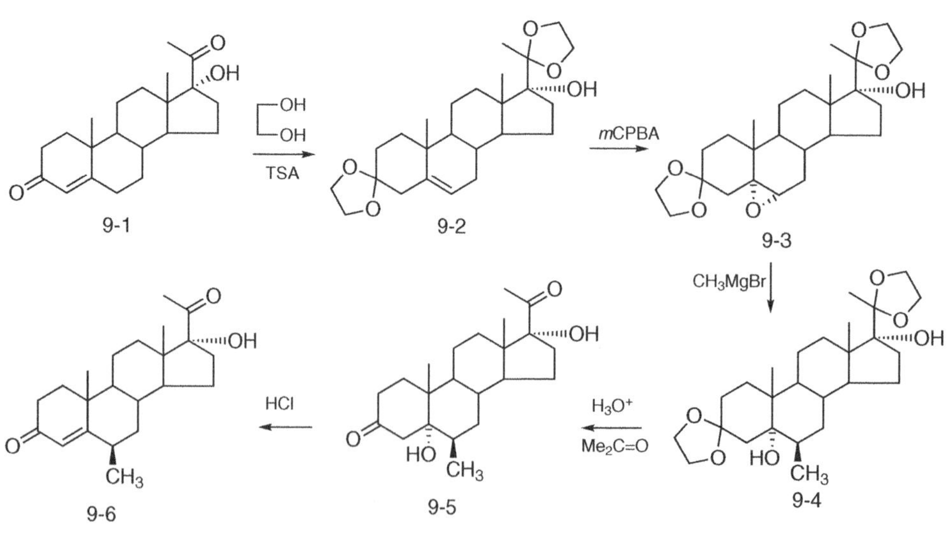

Scheme 6.9

acetal, often called a ketal, will feature prominently in the reaction schemes that follow. It is usually prepared by heating the carbonyl-containing compound and ethylene glycol and a catalytic amount of acid in a solvent such as benzene in a flask equipped with a trap to collect the ensuing water–benzene azeotrope. This transformation usually, as in the present example, causes the olefin to migrate from C_4 to C_5 (**9-2**). Reaction with a peracid, in this case *m*-chloroperbenzoic acid (*m*CPBA), leads to a mixture of peroxides in which the α-isomer predominates. Treatment of the major isomer with methylmagnesium bromide then opens the ring to form the 5-hydroxy-3-keto compound **9-5**. Strong acid leads the latter to dehydrate; this re-establishes the enone, now substituted, as above, with a 6β-methyl group (**9-6**).

The new methyl group can be epimerized by virtue of its location on an enolizable methylene group. Treatment with either strong acid or base leads the methyl group to epimerize to the favored, equatorial α-isomer **10-1** (Scheme 6.10). Acetylation under forcing conditions then gives the 17-acetate **10-2**. This product is the potent, widely used, orally active progestin medroxyprogesterone acetate, better known as Provera. Treatment of this with chloranil extends the conjugation of the ketone by deleting hydrogen from C_{6-7}. This product, named megestrol acetate, is yet another potent orally active progestin.

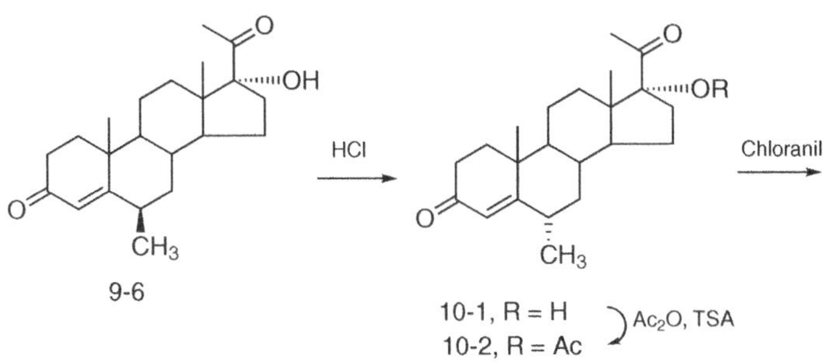

Scheme 6.10

An alternative method for introducing carbon at C_6 draws on a reaction more commonly used to modify aromatic rings. Reaction of 17-acetoxyprogesterone (**5-6**) with methyl orthoformate forms the enol ether **11-1** (Scheme 6.11). Treatment of this reactive intermediate with Vilsmeyer reagent (DMF and phosgene) introduces a methyleniminium group at C_6. Catalytic hydrogenation over Raney nickel reduces the iminium function to a methyl group with concomitant expulsion of nitrogen. Treatment with mineral acid then hydrolyzes the enol ether and epimerizes the new methyl group to give **10-2** Alternatively, reaction of the salt **11-2** with lithium borohydride selectively reduces the iminium function to give the dimethylaminomethylene derivative **11-3**. Exhaustive hydrolysis of the initial product **11-2** converts the iminium salt to an aldehyde and restores the enone system to give **11-4**.

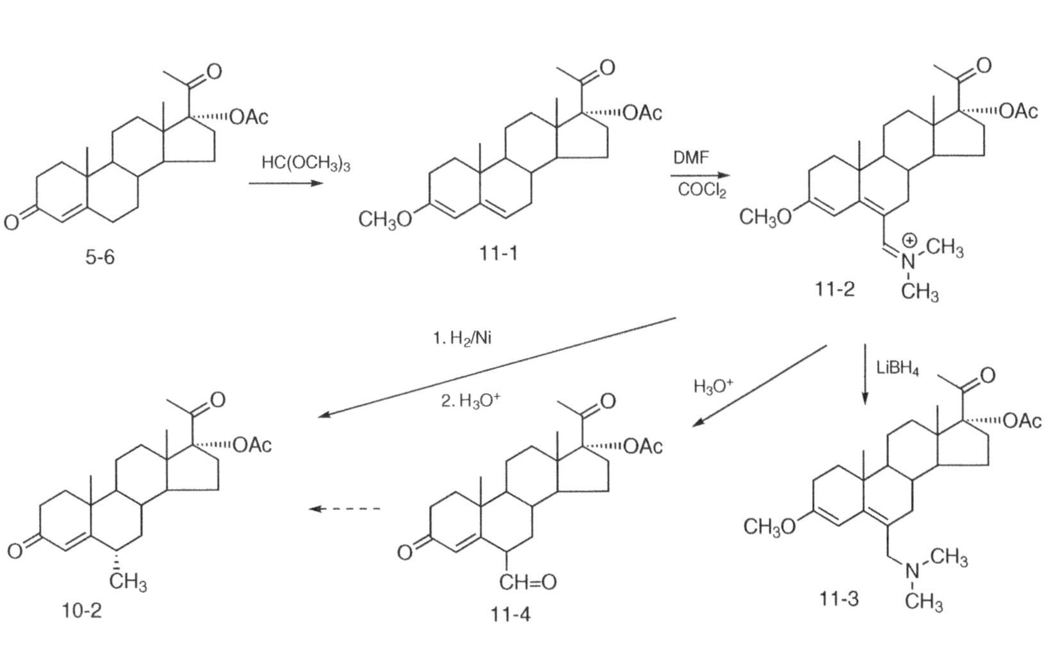

Scheme 6.11

Enol ether **11-1** also serves as starting material to another potent progestin that is orally active. Thus, treatment of **11-1** with *N*-chlorosuccinimide introduces halogen at the terminus of the diene system (**12-1**) (Scheme 6.12). This halogenation likely involves displacement of the succinimide by the electron-rich enolate. Acid hydrolysis restores the conjugated ketone (**12-2**); the stereochemistry of the product is of no concern since treatment of the intermediate with chloranil intended to introduce unsaturation at C_{6-7} makes that position trigonal. A second dehydrogenation step, this time using selenium dioxide, introduces a double bond at C_{1-2}. The product, chlormadinone (**12-4**), is used as an oral contraceptive either with or without added estrogen.

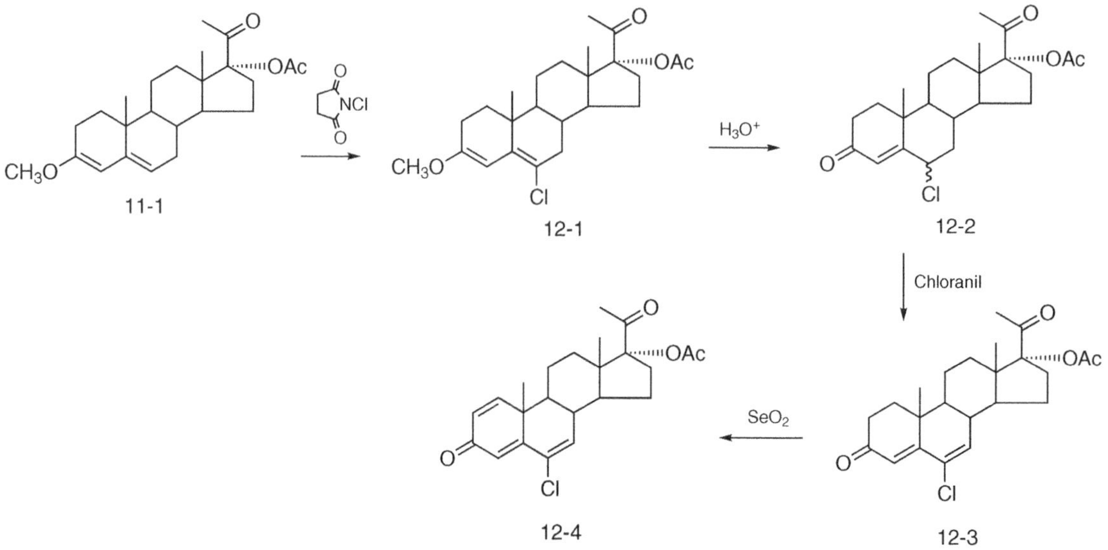

Scheme 6.12

Much the same sequence is used to prepare an analogue of **12-4** that also includes a cyclopropyl ring fused on to ring A. The synthesis of this compound starts with treatment of 1-acetoxyprogesterone with chloranil so as to establish the 4,6-dien-3-one function. Heating the product with selenium dioxide removes additional hydrogen, this time from ring A to form the 1,4,6-triene **13-1** (Scheme 6.13). Addition of diazomethane results in 1,3-cycloaddition of the reagent from the more open backside of the molecule to afford the steroid with a fused α-pyrrazoline (**13-2**). Pyrolysis of this product leads to extrusion of nitrogen with concomitant formation of the cyclopropyl ring with retention of the α-configuration (**13-3**). Reaction of **13-3** with perbenzoic acid leads to oxidation of the terminal olefin of the 4,6-dienone system (**13-4**). Exposure to hydrogen chloride in this case opens both the epoxide and the strained cyclopropyl ring to give the transient chlorohydrin **13-5**.

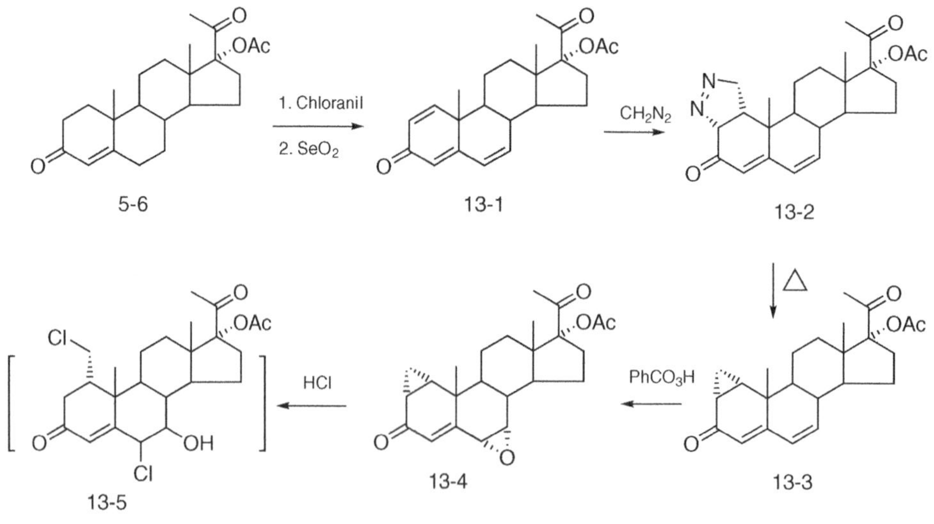

Scheme 6.13

The transient intermediate **13-5** is not isolated as it dehydrates spontaneously to give the intermediate **14-1** under the strongly acidic reaction conditions (Scheme 6.14). Treatment of that product with the hindered base collidine leads to formation of an anion on C_2. This displaces the side chain chlorine to re-form the cyclopropyl ring, yielding the progestin cyproterone (**14-2**).

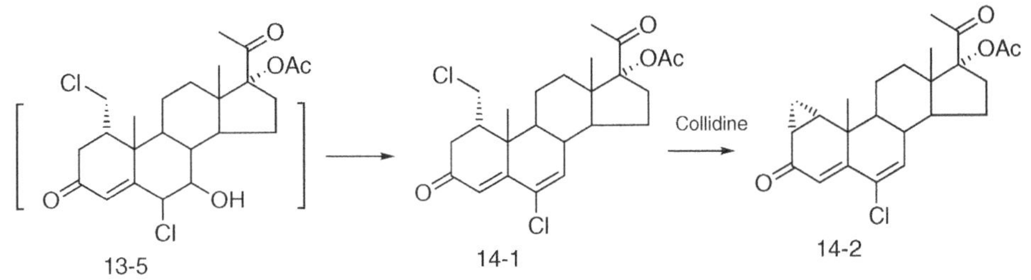

Scheme 6.14

The ethyl enol ether **15-1**, available from reaction of 17-hydroxyprogesterone with ethyl orthoformate, provides yet another starting material for medroxyprogesterone acetate. Thus, reaction of the enol ether **14-1** with carbon tetrabromide and pyridine leads to the bromomethylene intermediate **15-2** (Scheme 6.15). This reaction also likely starts by attack of the enolate on carbon tetrabromide to form a new carbon–carbon bond (for a related transformation, see Chapter 5, Scheme 6.27). Dehydrohalogenation of the initial transient bromomethyl derivative by pyridine in the reaction medium then leaves behind the *exo*-methylene bond (**15-2**). Catalytic reduction over strontium carbonate gives **15-2**.

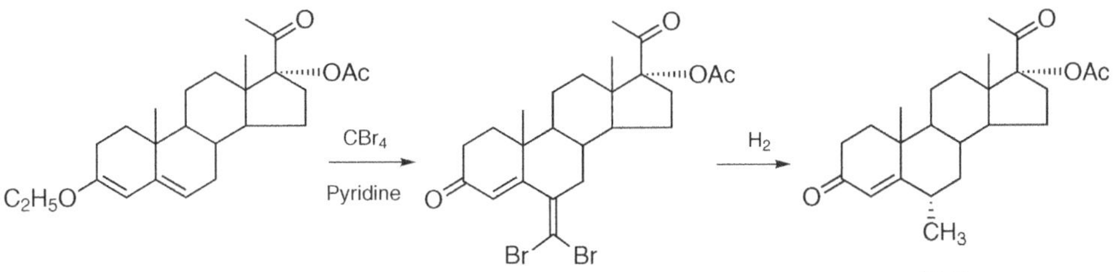

Scheme 6.15

A 6-methylprogesterone derivative lacking oxygen at C_3 exhibits the same biological activity as the parent drug; the same holds true in the 19-nor series, as noted previously (see Chapter 4, Scheme 4.22).

As in the 19-nor series, reaction of 17-hydroxy-6α-methylprogesterone (**10-1**) with ethylene dithioglycol proceeds to form the dithioacetal **16-1** (Scheme 6.16). Desulfurization by means of Raney nickel then yields the derivative with a methylene group instead of a ketone at C_3 (**16-2**). Treatment of that intermediate with acetic anhydride in the presence of *p*-toluenesulfonic acid gives the progestin angesterone (**16-3**).

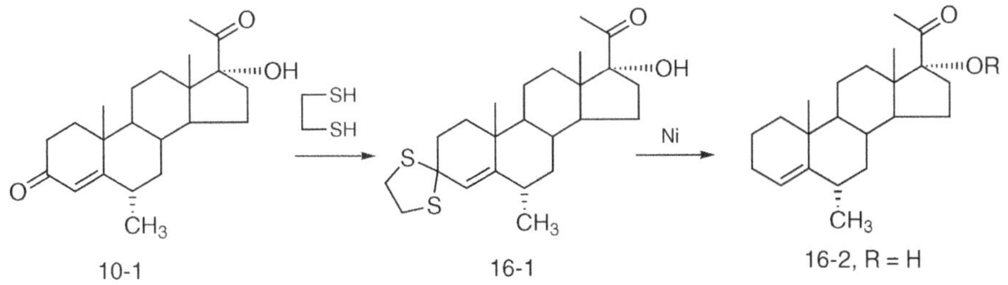

Scheme 6.16

6.3.4 General Methods for Modifications on Ring D

6.3.4.1 16α-Methyl Compounds

Many of the transformations used to modify progesterone, as noted in Chapter 7, are also applicable to corticosteroids. This applies particularly to the methods for introducing methyl groups at C_{16} in ring D. One scheme for adding a 16α-methyl group to these pregnanes involves conjugate addition of methylmagnesium bromide. In the example at hand, treatment of the acylated 16-dehydropregnenolone **17-1** with methylmagnesium bromide leads to the 16-methylated product **17-2** (the acyl group is lost by addition of the Grignard reagent) (Scheme 6.17). The α stereochemistry of the new methyl group is determined by addition of the organometallic reagent from the open α-side of the molecule.

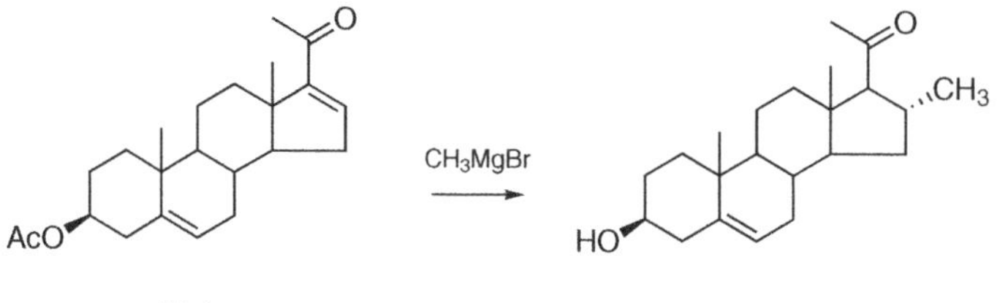

Scheme 6.17

Oppenauer oxidation of the intermediate **17-2** proceeds to give 16α-methylprogesterone (**18-1**) (Scheme 6.18). Selective introduction of the hydroxyl group at C_{17} depends on the differing reactivities of the two carbonyl groups. Reaction of the 3,20-diketone **18-1** with ethylene glycol proceeds to form an acetal selectively at C_3; additional hindrance the methyl group adjacent to the acetyl group sharpens the contrasting reactivity of the two ketones. Treatment of the product, **18-2**, with acetic anhydride, catalyzed by a strong acid such as perchloric acid, converts the side chain carbonyl to its enol acetate (**18-3**). Oxidation with a peracid the forms the C_{17-20} α-epoxide (**18-4**). This new functional array can be viewed as the acetate ester of an internal acetal where the latent hydroxyl at C_{17} acts as one of the acetal oxygens. Treatment with aqueous acid leads to hydrolysis of both that function and the ketal at C_3 to yield the progesterone derivative **18-5**.

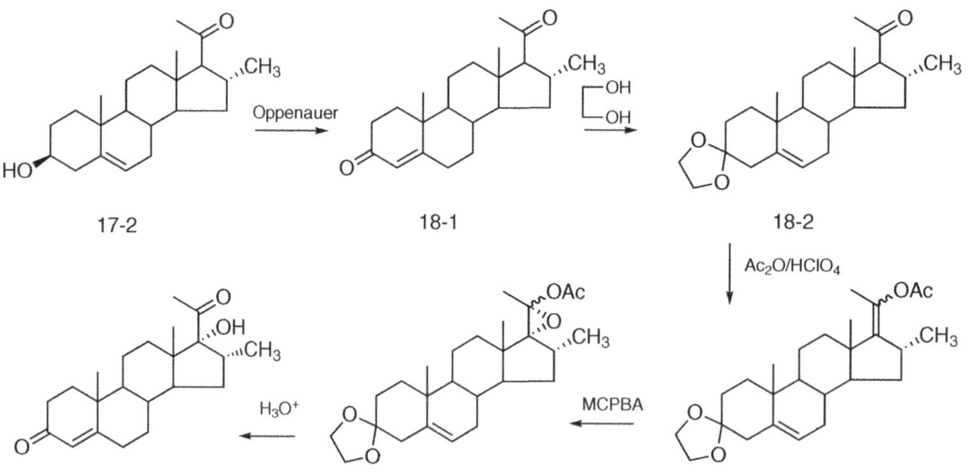

Scheme 6.18

6.3.4.2 16β-Methyl Compounds

Preparation of a progestin bearing a 16β-methyl group starts with 1,3-cycloaddition of diazomethane to the double bond in ring D to yield the fused pyrrazoline **19-1** (Scheme 6.19). Pyrolysis in this case yields the 16-methyl derivative **19-2**, in marked contrast to the adduct at position 1, which forms a fused cyclopropane (Scheme 6.13). Catalytic reduction adds hydrogen from the backside to afford the product **19-3**, which now features a 16β-methyl substituent. The fact that the double bond in ring B is also reduced will require additional steps to establish an unsaturated carbonyl group in ring A.

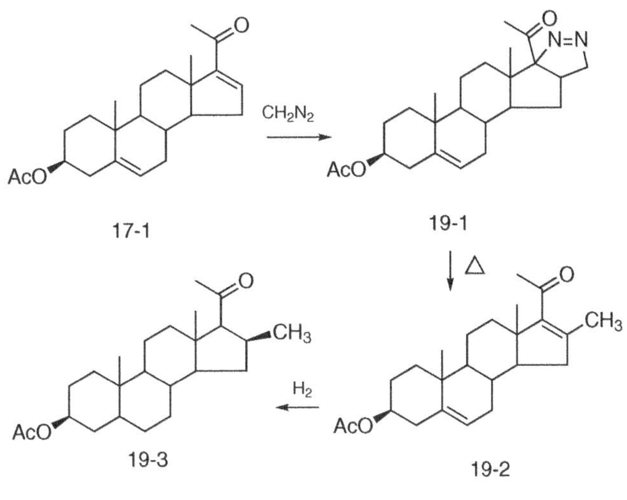

Scheme 6.19

6.3.4.3 Compounds with an Alkyl Group at C_{17}

Addition of an alkyl group at C_{17} initially involves conditions fairly similar to those used in the Birch reduction of enones. A solution of the steroid in liquid ammonia and some inert co-solvent is first treated with lithium metal (Scheme 6.20). The resulting anion **20-1** is then quenched with an alkyl halide instead of the alcohol usually employed in Birch reductions. The anion at C_{17} then displaces the halogen, in the case at hand iodide, from the alkyl group to form a carbon–carbon bond and the 17α-methyl derivative **20-2**.

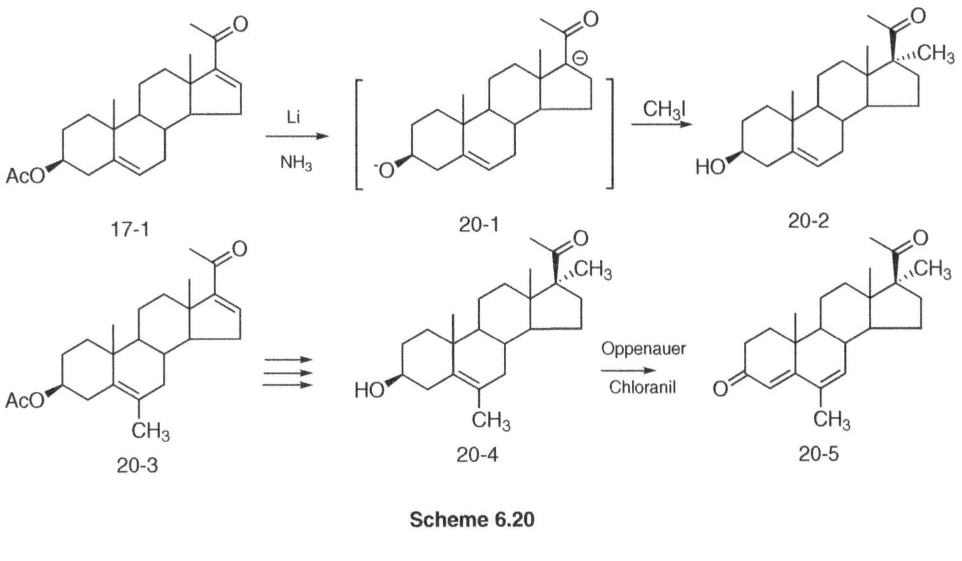

Scheme 6.20

The same scheme starting with 6-methyl-16-dehydropregnenolone acetate (**20-3**) leads to the 6,16-dimethyl analogue **20-4**. In an unusual variation of the Oppenauer oxidation, that product is treated in a one-pot reaction with aluminum isopropoxide and cyclohexanone together with chloranil. The latter apparently abstracts hydrogen from ring B as soon as the unsaturated ketone in ring B is formed. The product, medrogesterone (**20-5**), is another potent, orally active progestin.

6.3.4.4 16,17-Glycols

Preparation of a pregnane bearing hydroxyl groups at both C_{16} and C_{17} is relatively straightforward. The conservative approach to such a compound starts with acid-catalyzed reaction of 16-dehydroprogesterone (**1-4**) with ethylene glycol (Scheme 6.21). This leads to selective formation of a ketal from the ketone at C_3 in which the olefin has migrated to C_5 (**21-1**). Oxidation with osmium tetraoxide occurs exclusively at the conjugated double bond in ring D (**21-2**). The α-stereochemistry of the resulting diol is attributable to approach of the reagent from the backside. Aqueous acid the frees the carbonyl group in ring A (**21-3**). Reports in the literature indicate one-step direct oxidation of 16-dehydroprogesterone to the diol by means of potassium permanganate. Both the acetonide algestone acetonide (**21-4**), from reaction with acetone, and the acetophenide algestone acetophenide (**21-5**), from reaction with acetophenone, are potent progestins.

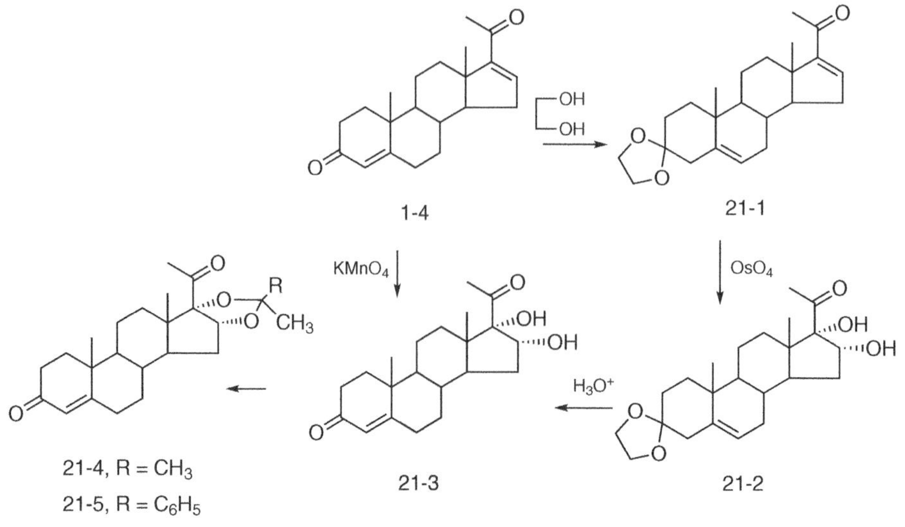

Scheme 6.21

6.3.5 More Progesterone Analogues

Many of the foregoing modified progesterone analogues exhibit improved biopharmaceutical properties over progesterone proper. The derivatives that follow were likely prepared in the hope that combinations of substituents would lead to additive improvements, a circumstance that often prevails in the pregnane series.

Acetylation of the 16α-methylpregnenalone derivative **17-2** leads to **22-1** (Scheme 6.22). Reaction of that derivative with chlorine leads to addition of the halogen to the double bond. The reaction in all likelihood first involves the formation of a chloronium ring between C_5 and C_6 from addition from the backside (**22-2**). The negative chloride counterion then opens the three-membered ring by front-side attack at C_6 to form a diaxial product **22-3**. Saponification of the acetate ester leads to the alcohol **22-4**. Oxidation of the newly freed hydroxyl group, for example with Jones' reagent (CrO_3 in acetone), leads to the diketone **22-5**. Dehydrochlorination is achieved by simply reacting **22-5** with sodium acetate. The stability of the conjugated ketone probably provides the driving force for this reaction, allowing the use of those mild conditions. The product from that reaction is the axial β-chloro analogue **22-6**. Treatment with dilute acid leads to rearrangement to the favored 6α-chloro equatorial epimer clometherone (**22-7**).

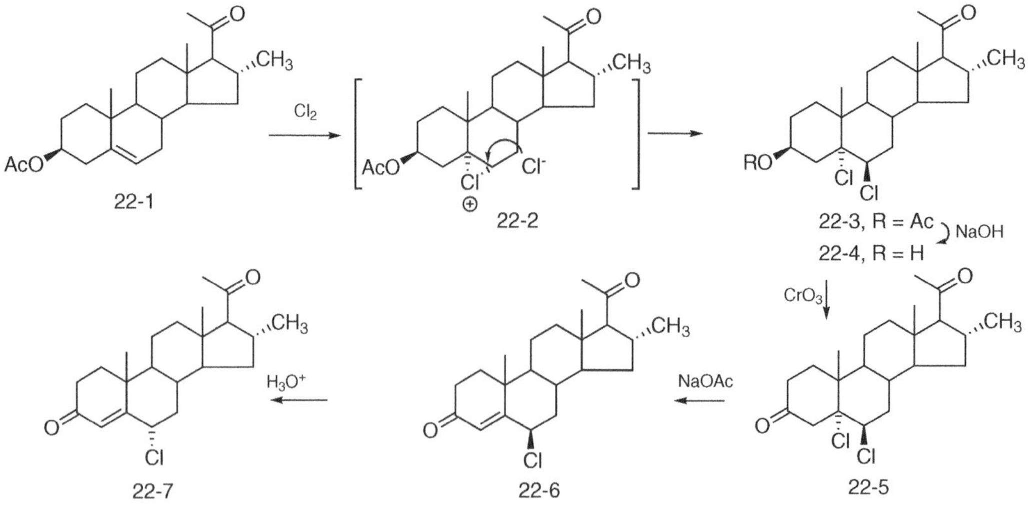

Scheme 6.22

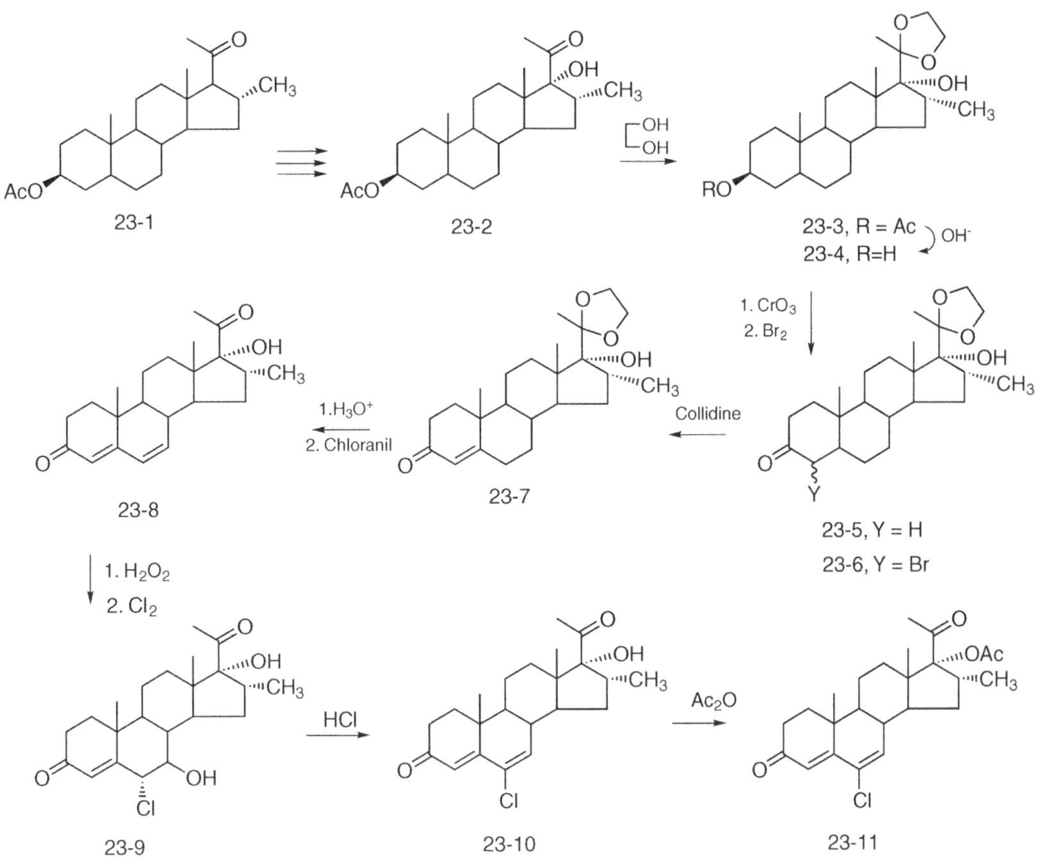

Scheme 6.23

A considerably more complex scheme is required for the preparation of the 17-oxygenated analogue of clometherone. Synthesis of that derivative essentially comprises carrying out in sequence the steps used to shape each of the individual modifications. The first of the subroutines involves adding oxygen at C_{17} by the same sequence as that discussed in Scheme 6.18. Intermediate **23-1** is thus converted to its enol acetate and that reacted with a peracid; hydrolysis of the resulting epoxide then affords the 17-hydroxyl derivative **23-2** (Scheme 6.23). The side chain ketone is next converted to its acetal with ethylene glycol (**23-3**) and the acetate at C_3 saponified (**23-4**). Oxidation of the newly revealed hydroxyl then gives the corresponding ketone **23-5**. Carefully controlled bromination adds a single halogen to give the 4-bromo derivative **23-6**. That derivative is next treated with a hindered base such as collidine; the resulting dehydrohalogenation inserts unsaturation at C_4; the conjugation is then extended by means of collidine to form the 4,6-diene **23-8**.

A sequence of several steps, first discussed in connection with Scheme 6.13, installs chlorine at C_6. Reaction of **23-8** with basic hydrogen peroxide starts by selective conjugate addition of peroxide anion to the distal end of the diene. That transient intermediate goes on to form the α-epoxide. Diaxial opening of the oxirane with chlorine gives chlorohydrin **23-9**. Mineral acid causes that product to dehydrate, in the process restoring the 4,6-diene **23-10**. Reaction with acetic anhydride under forcing conditions leads to acylation of the 17-hydroxyl group. That acetate, clomegestone (**23-11**), is somewhat more potent than its 17-desoxy congener clometherone (**22-7**).

Treatment of the 16-methylated pregnenolone **22-1** with *mc*-chloroperbenzoic acid leads to the corresponding α-epoxide **24-1** (Scheme 6.24). This function is stable to conditions in the subsequent four steps, serving as a protecting group for the double bond at C_{5-6}. Oxygen is next introduced at C_{17} by the same sequence as that discussed in Scheme 6.18. Application of those several steps starting from intermediate **24-1** affords the 17-hydroxyl derivative **24-4**. Reaction of this last intermediate with methylmagnesium bromide opens the epoxide that has been carried through the preceding reactions (**24-5**).

Oxidation of the resulting 3,5,17-triol, **24-5**, leads to conversion of the only secondary alcohol in this intermediate to a ketone (Scheme 6.25). This intermediate readily dehydrates to form the conjugated ketone present in most biologically active steroids (**25-1**). Treatment with chloranil then extends the conjugated ketone by a new double bond in ring B. The 17-acetate, **25-2**, formed by reaction with acetic anhydride under forcing conditions, is a potent progestin.

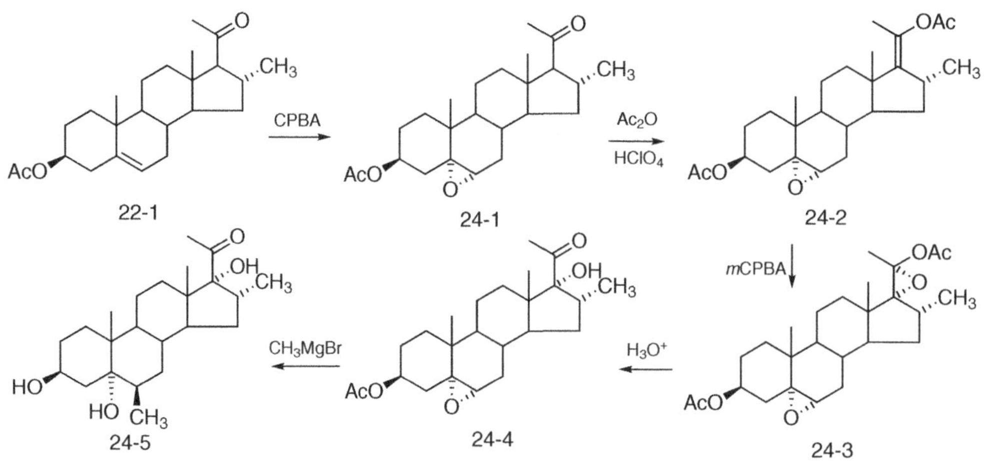

Scheme 6.24

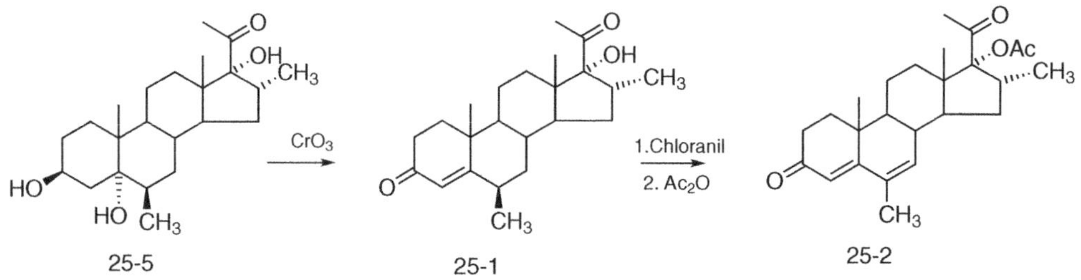

Scheme 6.25

The 6-methyl derivative of dehydropregnenolone acetate, **26-1**, whose synthesis is outlined in Chapter 2 (Scheme 2.10), comprises the starting material for yet another doubly modified progestin (Scheme 6.26). A methyl group is added at C_{16} by electrocyclic addition of diazomethane to that double bond, followed by pyrolytic elimination of nitrogen. As in the previous case (Scheme 6.19), the pyrolysis leads to the formation of a methyl group rather than a fused cyclopropane. Oxidation by means of alkaline hydrogen peroxide proceeds by initial conjugate addition of a peroxy anion, a process that obviously favors olefins conjugated with a carbonyl group. Reaction of the enone **26-2** with that reagent accordingly selectively converts the conjugated olefin in ring D to an epoxide (**26-3**). Both carbon atoms in the resulting α-oxirane are bonded to substituents on the opposite β-face of the molecule. This greatly hinders approach

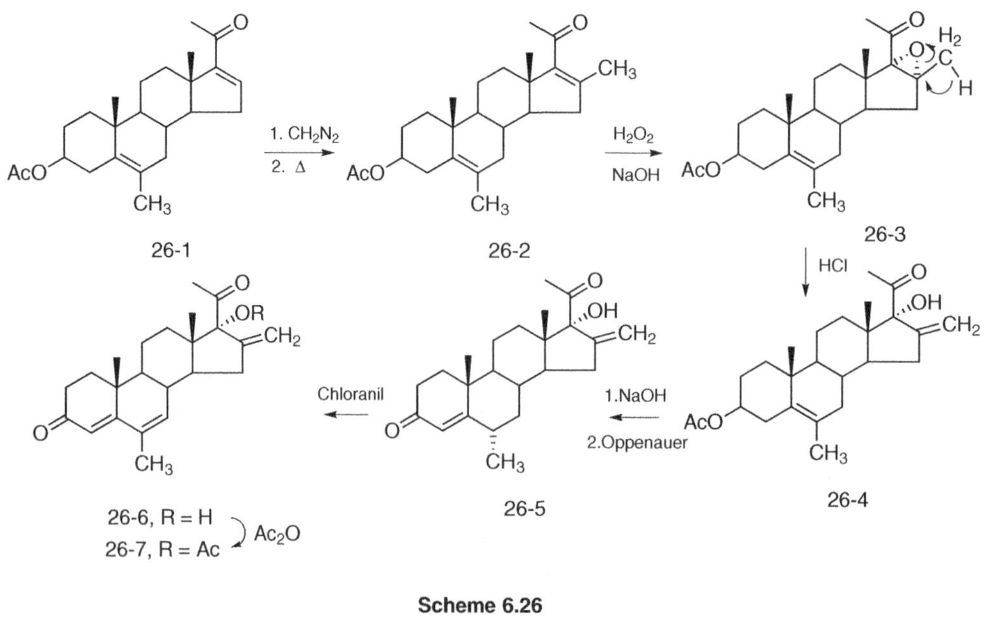

Scheme 6.26

of a reagent from the β-side, as is required for diaxial opening. Acid-catalyzed opening of the proximate epoxide consequently follows an unusual course. For book-keeping purposes, the net transformation involves the shift of a proton from the methyl on C_{16} to the adjacent carbon atom, opening the epoxide (**26-4**). Saponification of the acetate at C_3 followed by Oppenauer oxidation gives the enone **26-5**. That product is then dehydrogenated by means of chloranil (**26-6**) and the hydroxyl at C_{17} is acylated with acetic anhydride under forcing conditions. The final product, melengestrol (**26-7**), is yet another very potent progestin.

Addition of elemental bromine to pregnenolone acetate (**27-1**) causes two separate reactions; the halogen adds to both ends of the acetyl side chain to give the 17,21-dibromide and also adds to the double bond in ring B. The stereochemistry at the latter center reflects the intermediacy of a bromonium ion (**27-2**) (Scheme 6.27). Treatment of this intermediate 5,6,17,21-tetrabromide with sodium iodide displaces bromine at C_{21} with iodine; the hindrance about bromine at C_{17} portects that bromine from S_N2 displacement. Iodide ion also leads to expulsion of the adjacent bromines in ring B, restoring the 5,6-double bond. Sodium bisulfite then reductively removes the side chain iodine and the acetate is saponified to give **27-4**. The by now familiar sequence involving epoxidation of the isolated double bond in ring B followed by addition of a nucleophile, in this case fluoride, give the fluorohydrin **27-5**. The hydroxyl group at C_3 is then oxidized to a ketone, for example with Jones' reagent. Mineral acid causes the hydroxy ketone to dehydrate to give the conjugated ketone **27-6** that retains the β-fluoride in the precursor. The strong acid in the reaction medium causes the halogen to flip over to the favored equatorial α-configuration (**27-7**).

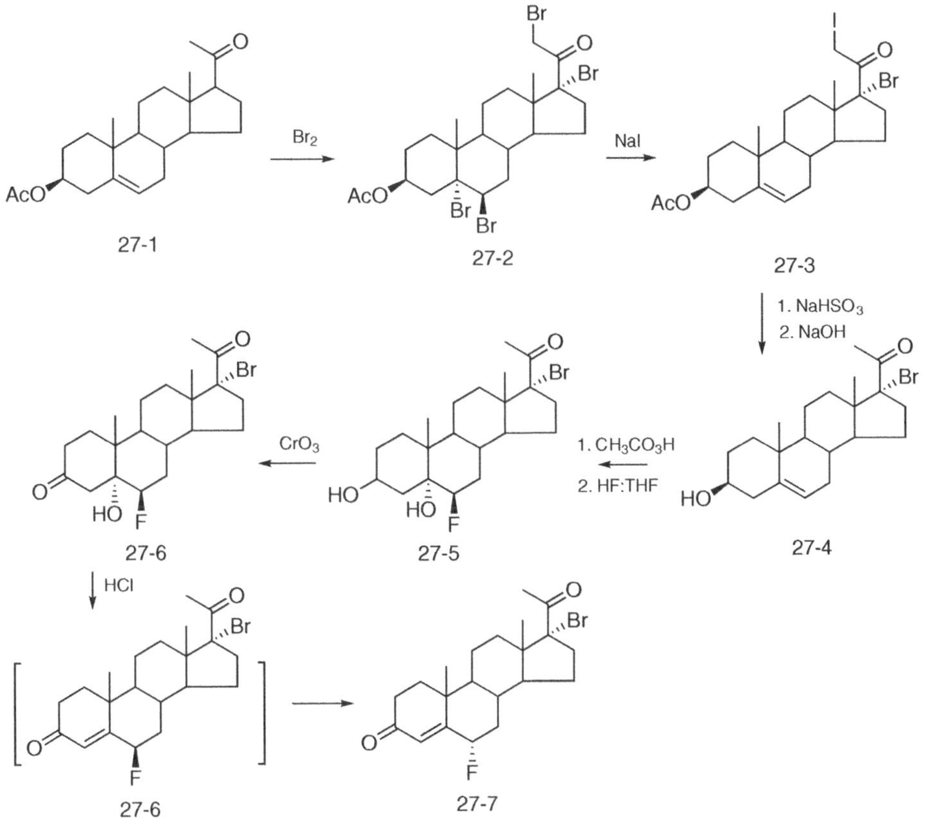

Scheme 6.27

16-Dehydropregnenolone acetate (**17-1**) serves as the starting material for a pregnane that bears methyl groups at both C_{16} and C_{17}. Both methyl groups are inserted in a one-pot reaction that comprises an interesting variant on a Grignard addition. Thus, reaction of the enone with methylmagnesium bromide catalyzed by cuprous ion leads to conjugate addition and formation of the enolate ion **28-1** (Scheme 6.28). Quenching that intermediate *in situ* with methyl iodide leads to alkylation of the ion and formation of **28-3**. Both steps of this reaction take place with approach of reagent from the more open backside, leading to the formation of the 16α,17α-dimethyl analogue **28-3**. Saponification, followed by Oppenauer oxidation, then gives 16α,17α-dimethylprogesterone (**28-4**).

The synthesis of a progestin receptor binding inhibitor based on a 19-nor retrosteroid is described in Chapter 4, Scheme 4.35. The corresponding analogue based on the progestin nucleus actually constitutes an approved commercial progestin. The preparation begins with bromination of the allylic 7-methylene group in pregnenolone acetate (**27-1**) by

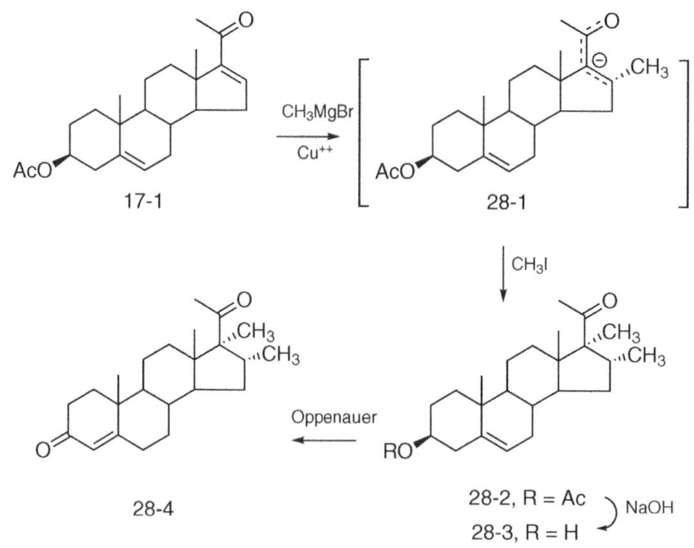

100 *Steroid Chemistry at a Glance*

Scheme 6.28

means of dibromodimethylhydantoin (DBDMH) (**29-1**) (Scheme 6.29). Dehydrobromination with collidine the generates new unsaturation at C_7. This bond array closely resembles that which occurs sin analogues of vitamin D (see Chapter 8, Scheme 8.24). Irradiation, as in the case of the vitamin, breaks the bond between C_9 and C_{10} (**29-4**). As this transient intermediate drops in energy it closes in conrotatory fashion to afford the so-called retrosteroid **29-5**, in which the stereochemistry at C_9 and C_{10} is reversed from that in normal pregnanes. Oppenauer reaction then oxidizes the hydroxyl with concomitant migration into conjugation of the proximate double bond. Treatment with acid cause the remaining double bond to migrate and also to give the 4,6-diene **29-6**. This product, dydrogesterone, is a progestin with an unusual pharmacological profile.

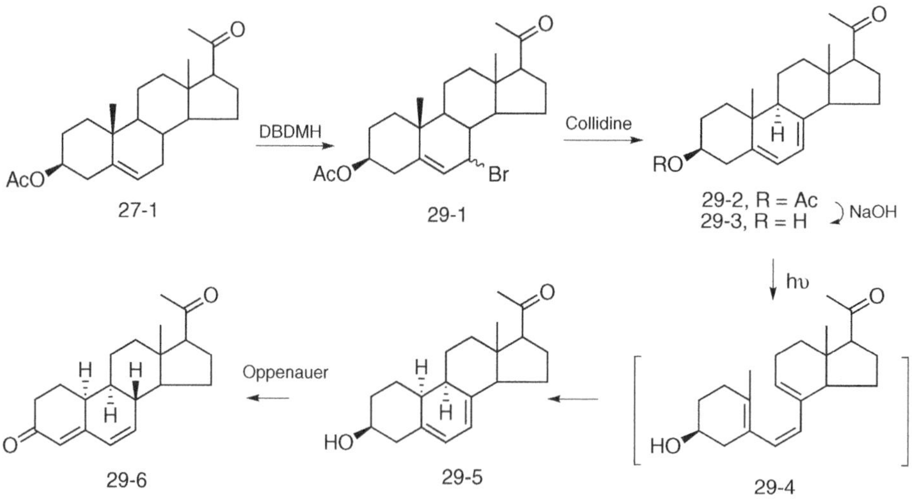

Scheme 6.29

6.4 Some Drugs Based on Progestins

6.4.1 Medroxyprogesterone Acetate (10-2)

The trade name of one, if not the earliest, orally active progestin to be approved for human use is Provera®, a name arguably constructed from *Pro* for progestin and *vera*, the Latin for true (the sponsor was arguably not aware that the name of one of Napoleon's generals was also Provera). This drug, today available in generic form, has found widespread use for treating conditions attributable to deficient progesterone levels. It also formed the progestin part of the former oral contraceptive Provest®. More recently, Pempro®, a fixed combination of medroxyprogesterone acetate and conjugated estrogens, was prescribed for treating menopausal women under the belief that this would be safer than the previously used pure estrogen. Adverse side-effects from a long-term study involving thousands of women led to an abrupt suspension of the trial well before its completion. An injectable solution of medroxyprogesterone acetate, named Depo-Provera®, provides a slow-release long-acting form of the drug. This form of the drug has found use as a contraceptive in those cases where oral contraceptives are contraindicated. According to sporadic reports in the press, injection of this formulation into male sex offenders, in order to blunt their sex drive, has been ordered by judges.

6.4.2 Megestrol Acetate (10-3)

The progestin megestrol acetate, which carried the trade name Megace® when first introduced, is now also available in generic form. Although the drug finds some use as an orally active progestin, it is mainly prescribed for treating cases of advanced breast and endometrial cancer. In addition to its antineoplastic activity, the drug improves the patients' desire for food. Loss of appetite, called cachexia, is a frequent and obviously undesirable effect from advanced malignancies and also AIDS.

6.4.3 Melengestrol Acetate (26-7)

By way of contrast, use of this potent and, again, orally active progestin is restricted to animals and then generally cattle. As a result of the suppression of gonadotrophins caused by the drug, melengestrol acetate also prevents cattle from coming into heat. Melengestrol acetate is consequently frequently used to synchronize heifers so that they ovulate at the same time as a group, increasing the efficiency of artificial insemination. Administration of the drug to cows prior to shipment prevents them from coming into heat during the transfer; this minimizes their potential loss of weight from the resulting agitated behavior.

7
Pregnanes, Part 2: Corticosteroids

7.1 Biological Activity

Addition of an oxygen atom to 17α-hydroxyprogesterone at C_{11} leads to a major qualitative change in the biological activity of the resulting molecule. One of the commonly used names for that group of compounds, corticosteroids, often contracted to corticoids, reflects their origin in the outer layer of the adrenal gland, the cortex. One subdivision of this steroid class comprises the glucocorticoids, which, as the name indicates, control mainly carbohydrate, protein and fat metabolism. Aldosterone, the principal mineralocorticoid steroid, maintains electrolyte and water balance. (Spirobutyrolactone aldosterone antagonists are discussed in the last section in Chapter 5.) This chapter focuses on steroids that will interact with glucocorticoid receptors. The serendipitous discovery by Kendall and Hench, in the late 1940s, of the anti-inflammatory activity of endogenous cortisone provided the impetus for research on the synthesis of cortisone from readily available starting materials. Once that had been accomplished, research shifted to the preparation of structurally related steroids. The synthesis programs that stemmed from that finding resulted in the preparation of very large collections of cortisone analogues. The biological activity of these steroids exhibited varying combinations of glucocorticoid and mineralocorticoid activity that reflect structural changes. The principal goal of each laboratory's synthesis program was the discovery of an analogue of cortisone that was more potent and that caused fewer side-effects. Many of the resulting molecules handily met the first target; the second was only partly achieved since many so-called side-effects were in fact actually extensions of the intrinsic biological activity of corticoids.

Many of the symptoms of inflammation are mediated by a set of molecules, named prostaglandins (PGs), derived from arachidonic acid, a 20 carbon tetra-unsaturated fatty acid. Non-steroid anti-inflammatory agents (NSAIDs) owe their efficacy to inhibition of the conversion of that acid to PGs. Corticoids on the other hand act one step earlier by inhibiting the action of the enzyme phospholipase A that liberates arachidonic acid from its stores in fatty tissues. Corticoids were at one time widely prescribed because of their sometimes dramatic reduction of inflammation. Serious hormonal effects that showed up on long term use has led to more limited indications. The drugs now tend to be used on an acute basis for treating threatening cases of inflammation that does not respond to NSAIDs. Specific corticoids find considerable use as anti-allergic agents particularly when applied topically. Longer term use includes replacement therapy for patients with primary adrenal insufficiency, also known as Addison's disease, as well as an adjunct to frank immunosupressors in connection with tissue or organ transplantation.

7.2 Sources of Corticoids

7.2.1 Introduction of Oxygen at C_{11}

7.2.1.1 Transposition (Bile Acid)

The apparent absence of any plant or animal product that contains oxygen at C_{11} poses a major stumbling block in any scheme for synthesizing corticoids. Cholic acid, abundantly available from slaughterhouses, was at one time considered as a source for corticoids. A sequence for transposing a hydroxyl group from C_{12} to C_{11} formed a key step in a very lengthy synthesis of cortisone from cholic acid. The first step involves heating the cholic acid derivative **1-1** at 320 °C under reduced pressure to give derivative **1-2**, that now includes a double bond in ring C (Scheme 7.1). Reaction of this intermediate with aqueous *N*-bromoacetamide (functionally HOBr) gave the bromohydrin **1-3** and also a C_{11}–C_{12} dibromo byproduct. One approach for proceeding from this point involves oxidation of the newly introduced hydroxyl group to a carbonyl group (**1-4**). Treatment with zinc then removes halogen to afford the 3,11-dione **1-5**. The overall yield for the sequence was, however, modest.

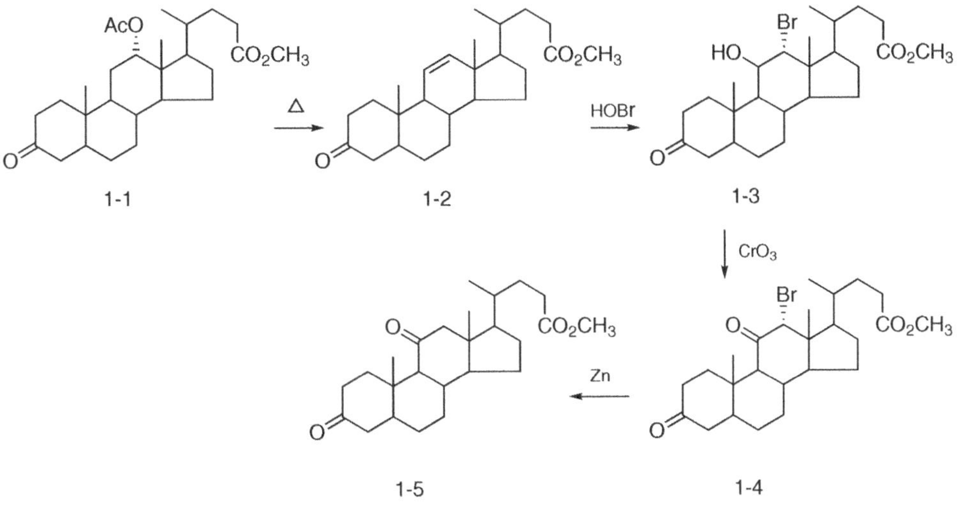

7.2.1.2 Fermentation

The real breakthrough for obtaining 11-oxygenated steroids came from a systematic program at the now-defunct Upjohn Company aimed at finding some microorganism that would oxidize the methylene group at C_{11}. In 1952, Peterson and his colleague Murray at Upjohn announced that they had successfully introduced a hydroxyl group into progesterone (**2-1**) by fermentation of the compound with a variety of *Rhizopus* actinomycete species. The product of this bio-oxidation was the hitherto unknown 11α-hydroxyprogeterone **2-2**; note that the α orientation of the key hydroxyl is opposite of that in corticoids. Oxidation converted that to the known pregnane-3,11,20-trione **2-3** (Scheme 7.2).

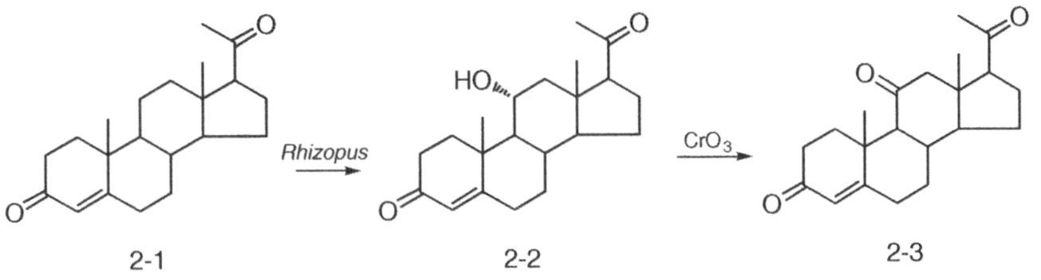

7.2.2 Construction of the Dihydroxyacetone Side Chain

7.2.2.1 Favorsky Chemistry

One scheme for converting the acetyl side chain to the corresponding 17,21-dihydroxy derivative found in most corticoids involves initial conversion of that unit to an unsaturated ester in which the carbonyl group is transposed to C_{21}. The process requires, first, selective bromination of the side chain methyl group in **2-3**. In order to insure selectivity, the methyl group at C_{21} is first activated by condensation with ethyl oxalate and sodium ethoxide. Reaction of the resulting sodium enolate **3-1** with bromine gives the 21-dibromo derivative, **3-2**; the oxalate fragment is expelled in the process by a reverse aldol reaction (Scheme 7.3). Treatment of this last intermediate with sodium methoxide now generates an anion at C_{17} (**3-3**). The charge on at C_{17} then displaces one of the bromines internally to form the transient cyclopropanone **3-4** characteristic of a Favorsky rearrangement. A second methoxide anion then adds to the strained carbonyl group, causing the ring to open so as to form the acrylate **3-5**.

The next sequence involves conversion of the acrylate side chain into the requisite dihydroxyacetone. Reaction of **3-5** with ethylene glycol catalyzed by a small amount of acid converts the carbonyl group at C_3 into the corresponding ethylene ketal **4-1**, protecting that group from the next reaction; the double bond shifts, as is customary, into ring B (Scheme 7.4). This reaction illustrates the very low reactivity of the ketone at C_{11}, particularly towards reagents that would create a bulky fragment, such as a cyclic acetal, at that position. Treatment of **4-1** with lithium aluminum hydride

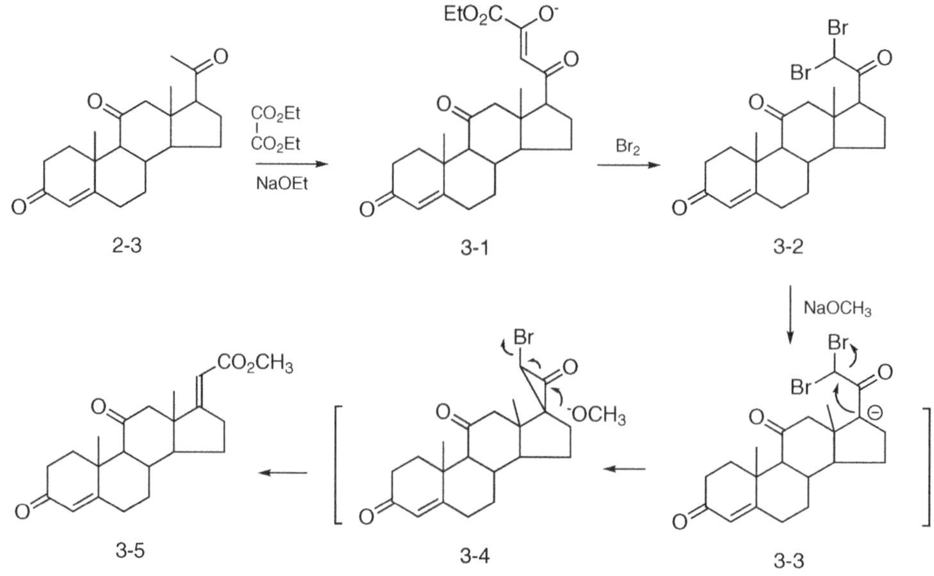

104 Steroid Chemistry at a Glance

Scheme 7.3

then converts the terminal ester into the desired hydroxymethylene, **4-2**. Hydride reducing agents comprise one of the few reagents to which the ketone at C_{11} is susceptible. That carbonyl group is thus reduced to the corresponding β-hydroxy derivative in the process. The side chain alcohol is then acylated by reaction with acetic anhydride to give the 21-acetate. Treatment with acetone in the presence of aqueous acid hydrolyzes the acetal to restore the carbonyl group at C_3. It remains to install an α-hydroxy ketone in the side chain. Osmium tetraoxide is a well recognized reagent for converting olefins to 1,2-glycols by way of an intermediate cyclic osmate ester. In order to spare this expensive and toxic reagent, it is used in far less than stoichiometric amounts. The co-oxidant, N-methylmorpholine oxide peroxide (NMOP), restores the osmium tetraoxide as it is used up. The secondary alcohol in the initial glycol **4-3** is then oxidized to a ketone by NMOP to give hydrocortisone acetate (**4-4**), also known as cortisol acetate. Oxidation of the alcohol at C_{11}, for example with Jones' reagent, then gives cortisone acetate (**4-5**).

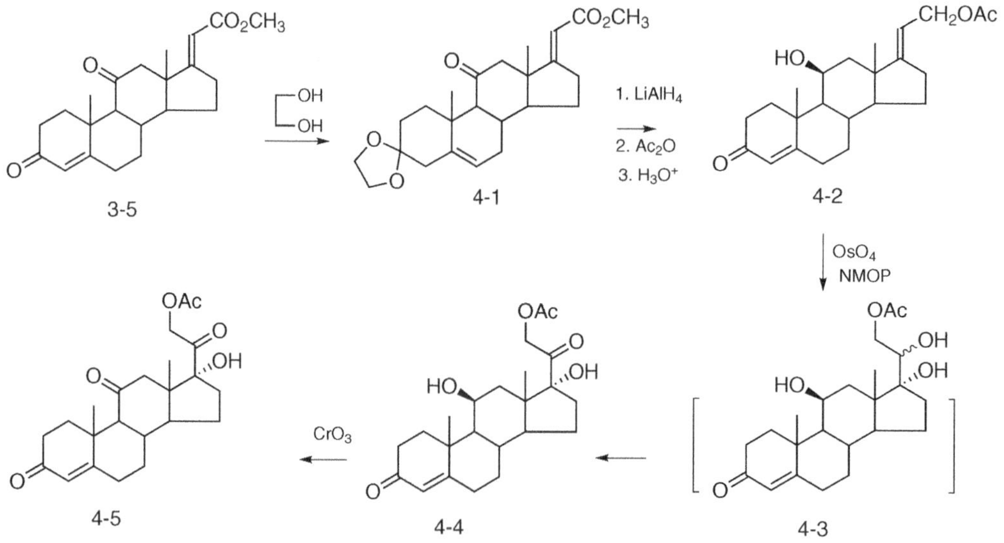

Scheme 7.4

7.2.2.2 Gallagher Chemistry

The first few steps in an alternative scheme are analogous to those used to introduce a hydroxyl group at C_{17} in progestins in Chapter 6 (Scheme 6.18). Treatment of the 11-oxopregnane (**5-1**) with acetic anhydride in the presence of mineral acids converts both ketones to their respective enol acetates; the hydroxyl at C_3 is also acetylated in the process (**5-2**) (Scheme 7.5). Oxidation of that intermediate converts the side chain unsaturation to an epoxide; the enol ester in ring C is unaffected (**5-3**). The newly formed side chain function in effect comprises an internal acetal. Mild saponification removes all three acetate groups: the epoxy-acetate at C_{17-20} collapses to a hydroxy ketone while the enol acetate in ring C reverts to a ketone. The acetate at C_3 also comes off leave a free hydroxyl group (**5-4**). Reaction with bromine places one bromine on the side chain methyl group (**5-5**). This is then displaced by means of sodium acetate to give the desired dihydroxy ketone side chain as its 21-acetate. This intermediate has been taken on to cortisone acetate (**5-7**) in a few steps.

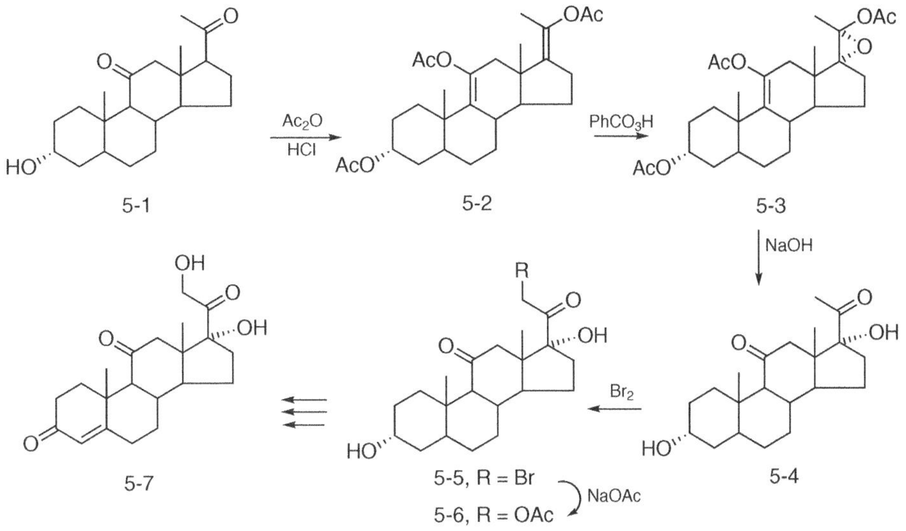

Scheme 7.5

7.3 Modified Corticoids

It becomes increasingly difficult to classify modification of the structures of steroids in neat sections as the chemical structures of corticoids become more complex. In the case of the corticoids, increases in potency due specific modifications tend to be additive. Many corticoids consequently incorporate several such diverse modifications in a single molecule. The classification that follows is therefore somewhat arbitrary.

7.3.1 Unsaturation

Treatment of a corticoid that bears the 4-en-3-one function with selenium dioxide abstracts hydrogen from ring A to introduce a new double bond to form a 1,4-dien-3-one. Application of this reagent to hydrocortisone acetate (**4-4**) results in the formation of prednisolone (**6-1**) (Scheme 7.6). The corresponding reaction on cortisone acetate (**4-5**) leads to prednisone acetate (**6-2**).

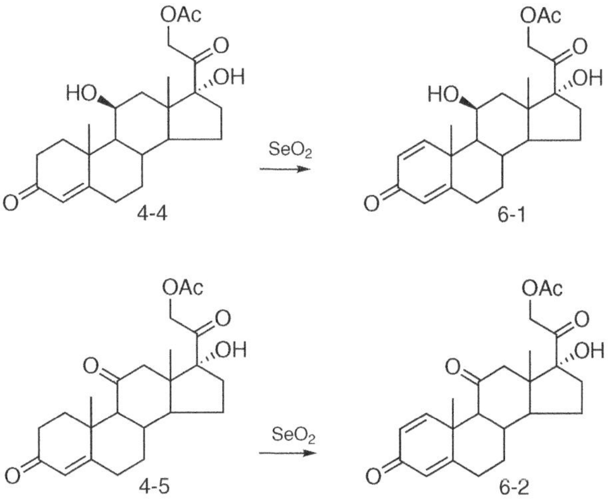

Scheme 7.6

7.3.2 Additional Alkyl Groups

The 17,21-dihydroxyacet-20-one side chain present in the majority of corticoids frequently needs to be protected in the course of some synthetic sequence. This is most conveniently accomplished by taking advantage of the juxtaposition of the two hydroxyl groups and the ketone in the C_{17} side chain. Formation of a specialized protecting group that takes advantage of that array of functional groups, dubbed bismethylenedioxy (BMD), involves an acid-catalyzed condensation reaction of the side chain with formaldehyde. The sequence can be visualized by assuming that the terminal hydroxyl on the generalized corticoid side chain (**7-1**) forms a hemiacetal (**7-2**) with formaldehyde (Scheme 7.7). The resulting hemiacetal hydroxyl then adds to the adjacent carbonyl group to form a transient acetal such as **7-3**. A second equivalent of formaldehyde now reacts with the hydroxyl group at C_{20} to form the new hemiacetal **7-4**. The final step comprises ring closure to **7-5**. This complex fragment is often depicted in structural formulae in the abbreviated form 'BMD' (**7-6**).

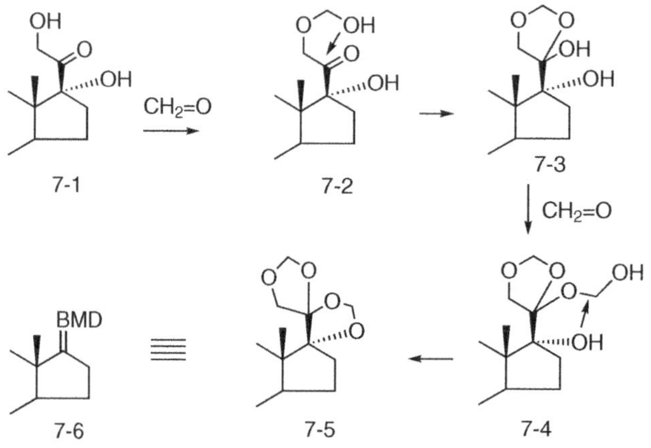

Scheme 7.7

Synthesis of a corticoid that will feature an additional methyl group at C_6 starts with the reaction of cortisone (**5-7**) with formaldehyde to convert the dihydroxyacetone side chain to its BMD derivative. Subsequent acid-catalyzed condensation of the product, **8-1**, with ethylene glycol serves two purposes: the carbonyl group in the product **8-2** is protected as its ethylene acetal and the reaction shifts the double bond into ring B, thus providing functionality for introducing the methyl group (Scheme 7.8). Oxidation of the shifted double bond in **8-2** with perbenzoic acid gives the epoxide **8-3** as a mixture of stereoisomers. Treatment with formic acid then causes the epoxide to rearrange to a carbonyl group (**8-4**). Addition of methylmagnesium bromide leads to addition of the organometallic to the carbonyl group at C_6 from the underside to give the hydroxymethyl derivative **8-5**. The other ketone in the compound, at C_{11}, is, as expected, inert to the Grignard reagent.

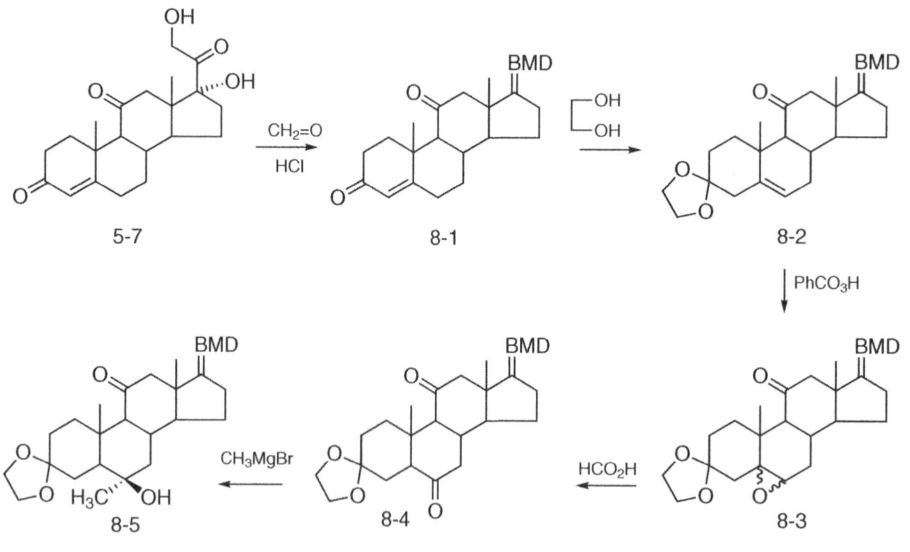

Scheme 7.8

The tertiary hydroxyl group dehydrates when **8-5** is treated with *p*-toluenesulfonic acid (TSA). The refractory carbonyl group at C_{11} will, however, react with most reducing agents. Lithium aluminum hydride thus reduces that ketone in **9-1** to form the β-hydroxyl derivative (Scheme 7.9). Hydrolysis of the acetal then frees the carbonyl group; the unsaturation moves into conjugation in the process. The acidic conditions in this last reaction cause the newly introduced methyl group to assume the favored equatorial, in this case α, configuration (**9-2**). Reaction with selenium dioxide abstracts hydrogen from ring A, establishing the now familiar 1,4-dien-3-one array. Exposure to acetic acid then hydrolyzes the BMD grouping to reveal the 17,21-dihydroxy-20-one side chain (**9-4**). The final product, methylprednisolone (**9-4**), is a widely prescribed corticoid.

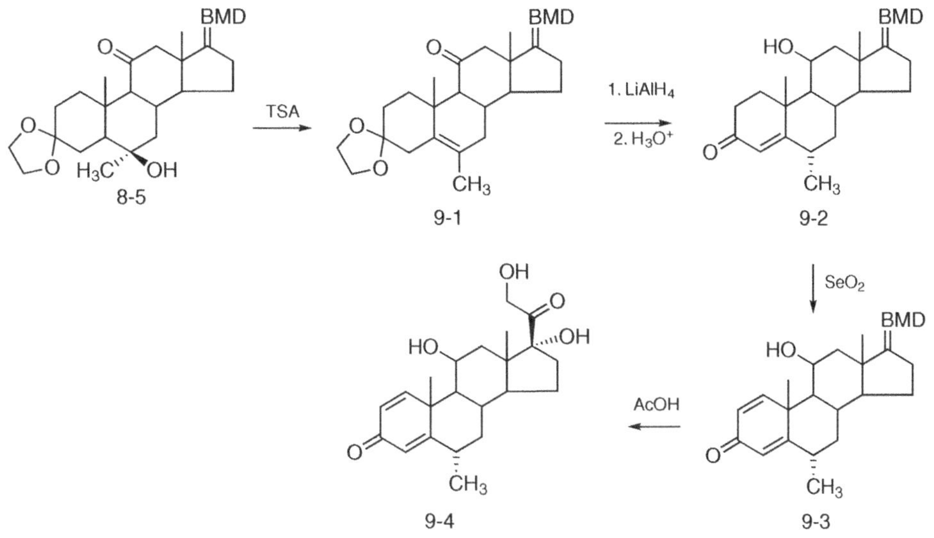

Scheme 7.9

A similar sequence is used to prepare a closely related corticoid that lacks the two side chain hydroxyl groups. The sequence for preparing this agent actually uses the actual fermentation product from progesterone, **2-2**, as starting material (see Scheme 7.2). The α-hydroxyl at C_{11} is much less hindered that the β-epimer and consequently more reactive. That alcohol is first converted to its benzoate ester **10-2**, for example by acylation with benzoyl chloride (Scheme 7.10). Peracid oxidation in this case affords predominantly the β-epoxide; the benzoate group may help steer the entering peracid. In a series of reactions analogous to that used to add methyl groups to

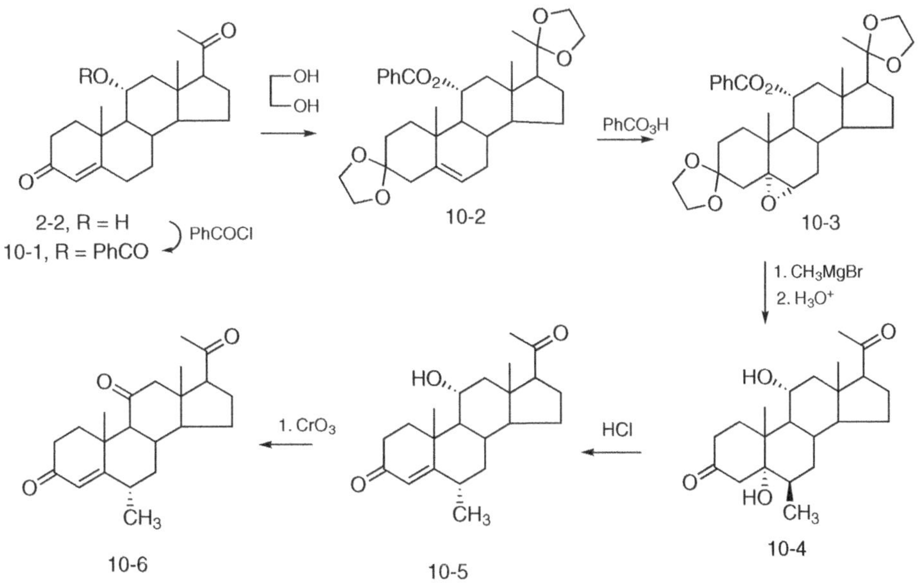

Scheme 7.10

pregnanes, epoxide **10-3** is first allowed to react with methylmagnesium bromide; axial opening yields the 6β-methyl-5α-hydroxy product, **10-4**. The 11-benzoate ester is cleaved in the process Exposure to mild acid hydrolyzes the acetals at C_3 and C_{20} to the corresponding ketones. Acid-catalyzed dehydration then causes the methyl group to be epimerized to the favored equatorial, α configuration (**10-5**). It remains to invert the hydroxyl group at C_{11} to the biologically active β configuration. The first step in this transformation comprises oxidation of that hydroxyl to the ketone (**10-6**).

Reaction of that triketone with ethylene glycol selectively forms the expected 3,20-bisacetal; the still free remaining ketone at C_{11} is next reduced with lithium aluminum hydride to yield the expected β-hydroxyl group (**11-1**) (Scheme 7.11). Acid hydrolysis of the acetal groups followed by reaction with selenium dioxide generates the 1,4-dien-3-one functional array. The product medrysone (**11-2**) is used mainly in anti-allergic eye drops.

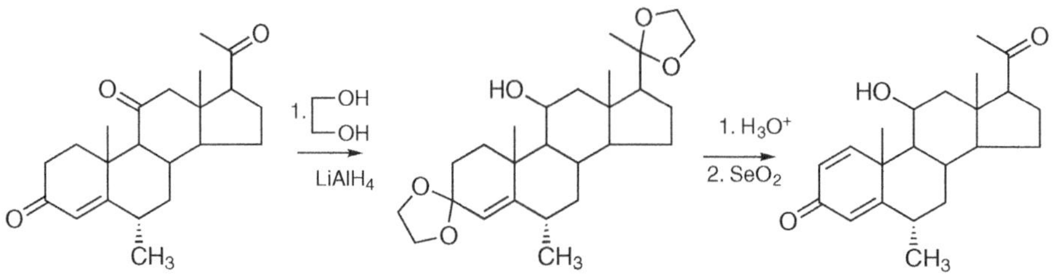

Scheme 7.11

One of the principal routes for metabolic inactivation of corticoids involves degradation of the dihydroxyacetone side chain. Substituents at C_{16} adjacent to that side chain slow inactivation by hindering the approach of the enzymes that degrade the dihydroxyacetone. One synthesis of a 16β-methylcorticoid starts with a late intermediate from the bile acid route. The requisite methyl group is added by first reacting **12-1** with diazomethane to give the pyrazole from 1,3-cycloadditon of the reagent (Scheme 7.12). Pyrolysis of that product extrudes nitrogen to afford the 16-methyl derivative **12-2** (see Chapter 6, Scheme 6.19). Catalytic reduction adds hydrogen from the backside to yield the 16β-methyl derivative **12-3**. The acetyl group at C_{17} is then converted to a dihydroxyacetone function by the Gallagher sequence discussed above (Scheme 7.5). Saponification of the product from that sequence, **12-4**, then cleaves the acetate esters. Oxidation of the resulting 3,17,21-triol with N-bromosuccinimide interestingly proceeds selectively on ring A to give the corresponding 3-ketone **12-5**. The selectivity of this step may be due to hydrogen bonding of the side chain hydroxyl with the adjacent ketone at C_{20}. Introduction of unsaturation in ring A draws on a scheme used in the preparation of estrone (Chapter 3, Scheme 3.2). Thus, bromination of **12-5** yields the

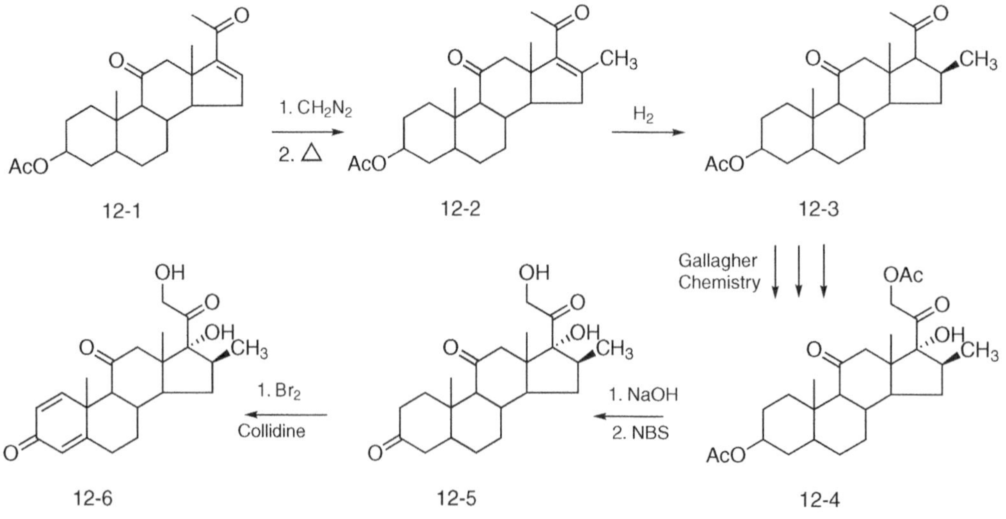

Scheme 7.12

corresponding 2,4-dibromide; dehydrobromination with a base such as collidine then affords the 1,4-dien-3-one, 16β-methylprednisone (**12-6**).

Starting material **13-1** for a corticoid that incorporates three additional methyl groups can in principle be obtained from **12-1** in several steps involving protection–deprotection sequences. The enolate at C_{21} from treatment of **13-1** with lithium diisopropylamide (LDA) is then alkylated with methyl iodide to afford an intermediate, **13-2**, in which the side chain on ring D has been extended by one carbon (Scheme 7.13).

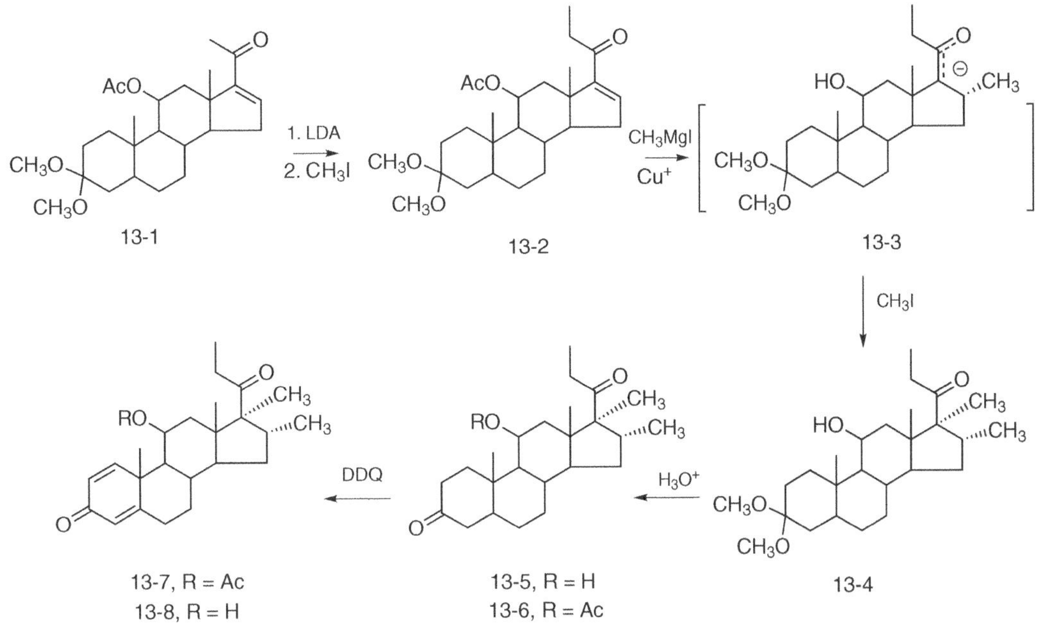

Scheme 7.13

In a sequence parallel to that used in the progestin series (see Chapter 6, Scheme 6.28), methylmagnesium bromide is added to the enone in ring D in the presence of cuprate ion. Addition takes place from the backside to form the enolate **13-3** of the 16α-methyl derivative. This charged species is then quenched with methyl iodide. This adds a second methyl, again from the backside, to give the 16α,17α-dimethyl derivative **13-4**; the acetyl ester at C_{11} is cleaved in the process. Treatment with mild acid then hydrolyzes the dimethyl acetal (**13-5**); the hydroxyl at C_{11} is next re-acetylated with acetic anhydride and strong acid to cover the hydroxyl in the subsequent step (**13-6**). Reaction of this last intermediate with the powerful dehydrogenation agent 2,3-dichloro-5,6-dicyanoquinone (DDQ) introduces the two double bonds in ring A to give the 1,4-dien-3-one rimexolone (**13-8**).

7.3.3 Halogenated Corticoids

The effect on the potency of steroids of adding the extra double in ring A has resulted in the incorporation of this structural feature in the majority of anti-inflammatory and anti-allergic corticoids. This observation also holds true for inclusion of halogen atoms and most frequently fluorine in the structure. Fluorine at C_9 and/or C_6 is thus found in the majority of corticoids. The very versatile sequence for introducing fluorine discussed in Chapter 5 (Scheme 5.31) was most likely first developed and subsequently applied in the corticoid series.

As in the previous example, dehydration in ring C, in this case by means of phosphorus oxychloride in pyridine, gives the Δ(9)11-olefin, **14-1** (Scheme 7.14). Treatment of the resulting intermediate with N-bromoacetamide, or some other source of electrophilic bromide ion, in water (in effect HOBr) leads to the formation of the bromohydrin **14-2**. The stereochemistry is dictated by the initial formation of a bridged α-bromonium ion from attack on the more open underside of the molecule. Axial opening by hydroxide then gives the observed bromohydrin (**14-2**).

Reaction with base causes the resulting 11-alkoxide to displace the adjacent bromine to form the β-oxirane **14-3**. Hydrogen fluoride in THF leads to attack on the oxirane by fluoride anion. Diaxial ring opening gives the 9α-fluoro-11β-hydroxy derivative fludrocortisone acetate (**14-4**). The corresponding 1,4-diene 9α-fluorprednisolone (**14-5**), is obtained by reaction of **14-4** with a dehydrogenation agent such as DDQ.

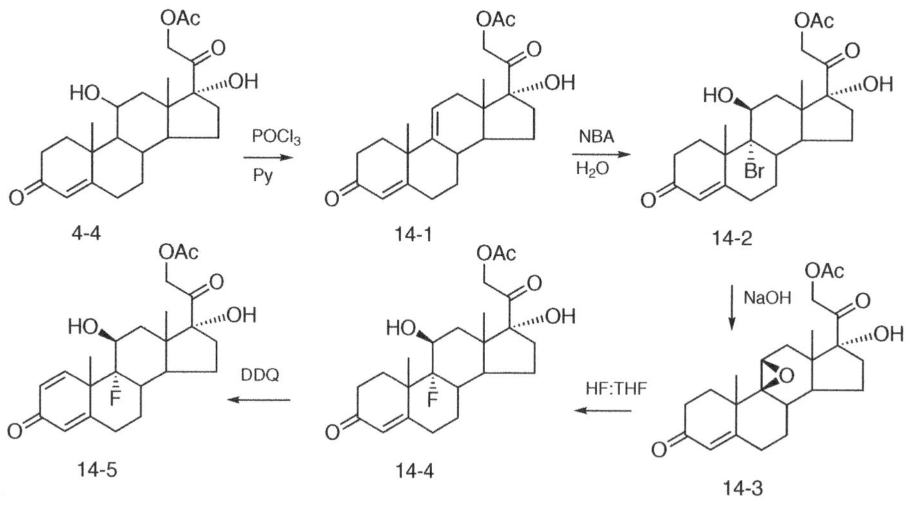

Scheme 7.14

Elemental chlorine can replace the hydrofluoric that was acid used to prepare the 9α-fluoro-10β-hydroxy array. The substrate for that addition, **15-1**, can be prepared by dehydrating prednisolone acetate (**6-1**) (Scheme 7.15). Reaction of the resulting 1,4,(9)11-triene with chlorine leads to the 9α,10β-dichloro corticoid dichlorisone (**15-2**). The rationale for the stereochemistry of this product invokes the intermediacy of a C_{9-11} α-bridged chloronium intermediate.

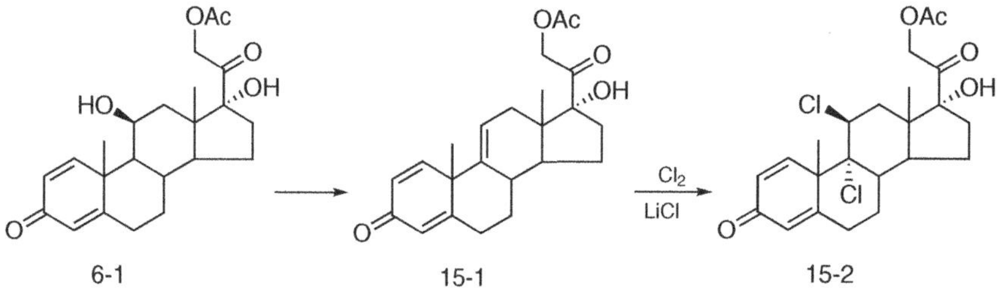

Scheme 7.15

The sequence for introducing halogen at C_6, specifically chlorine, closely follows that used to prepare 6-chloroprogestins (see Chapter 6, Scheme 6.12). The synthesis starts with the formation of the enol ether of cortisone acetate, **4-4**, by means of methyl orthoformate (**16-1**) (Scheme 7.16). Reaction of that intermediate with *N*-chloroacetamide gives the 6-chlorinated derivative. Treatment with mild acid hydrolyzes the enol ether to give **16-2** as a mixture of isomers that consists primarily of the 6β-chloro epimer. Strong acid converts the mixture to the single α-epimer that comprises

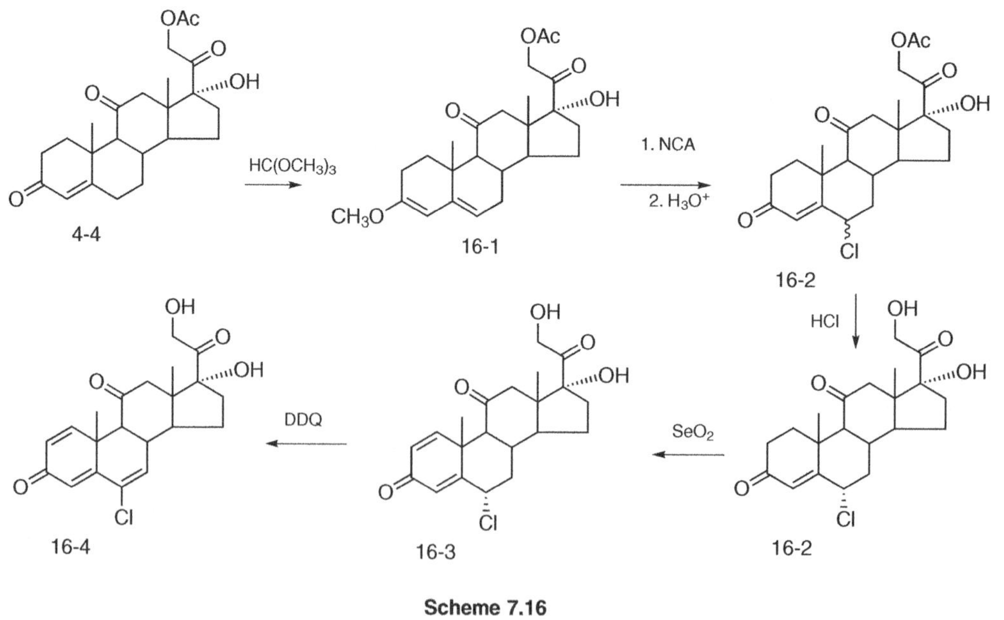

Scheme 7.16

the more stable, equatorial 6α-chloro derivative; the reaction medium cleaves the acetate on C_{21} to the corresponding free hydroxyl. An additional double bond is next introduced in ring A by means of selenium dioxide to give chloroprednisone (**16-3**). Further dehydrogenation, this time with DDQ, extends the conjugated system, yielding cloprednol (**16-4**).

Combining the strategy used to introduce halogen at C_9 with that for adding halogen at C_6 leads to the corresponding 6,9-dihalocorticoids. The sequence starts with the dehydration of hydrocortisone (**5-7**) to afford the 4,(9)10-diene **17-1** (Scheme 7.17). Exhaustive acetylation of that product with acetic anhydride under forcing conditions gives the 17,21-diacetate; the enol form of the carbonyl group at C_3 is also acetylated to give enol acetate **17-2**. The newly introduced double bond in ring B in effect functionalizes C_6 towards electrophilic attack by halogen as described above. Thus, reaction of that intermediate with perchloryl fluoride ($FClO_3$) leads to the corresponding 6-fluoro derivative. Work-up affords the crude 6-fluoro derivative as a mixture of isomers in which the axial 6β-fluoro epimer (**17-3**) predominates. Treatment with strong acid isomerizes the mixture, as in the previous examples, to the favored equatorial 6α-epimer (**17-4**). It now remains to insert another fluorine atom at C_9. Following the usual procedure, the latter intermediate is first converted to the bromohydrin with N-bromoacetamide or an equivalent N-halo derivative. The alkoxide from treatment of the halohydrin with base displaces bromine from the adjacent atom to form the (9)11β-oxirane **17-5**. Reaction with hydrogen fluoride in THF opens the epoxide to the observed halohydrin. The traditional additional double bond in ring A is then introduced by means of selenium dioxide to afford the 6α,9α-difluoro corticoid **17-6**.

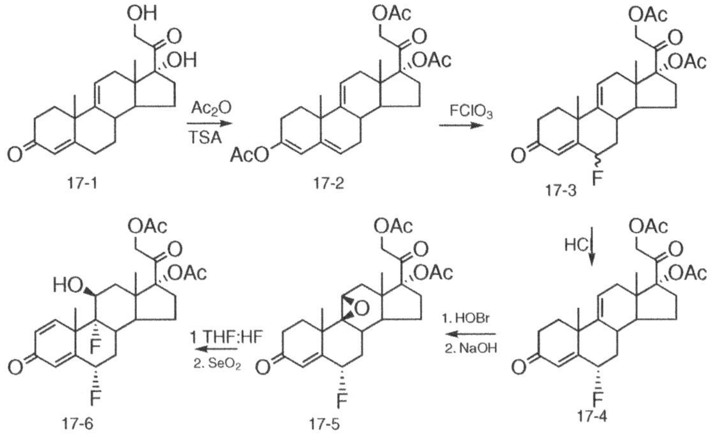

Scheme 7.17

7.3.4 Hydroxylation: 16,17-Diols

The presence of a hydroxyl group at C_{16} has an enhancing effect on potency similar to that of a methyl group at the same position. Cortisone (**5-7**) constitutes the starting material for the synthesis of one of those 16β,17β-diols. Dehydration of cortisone occurs in ring D to afford the corresponding 16,17-dehydro derivative **18-1** (Scheme 7.18). The carbonyl

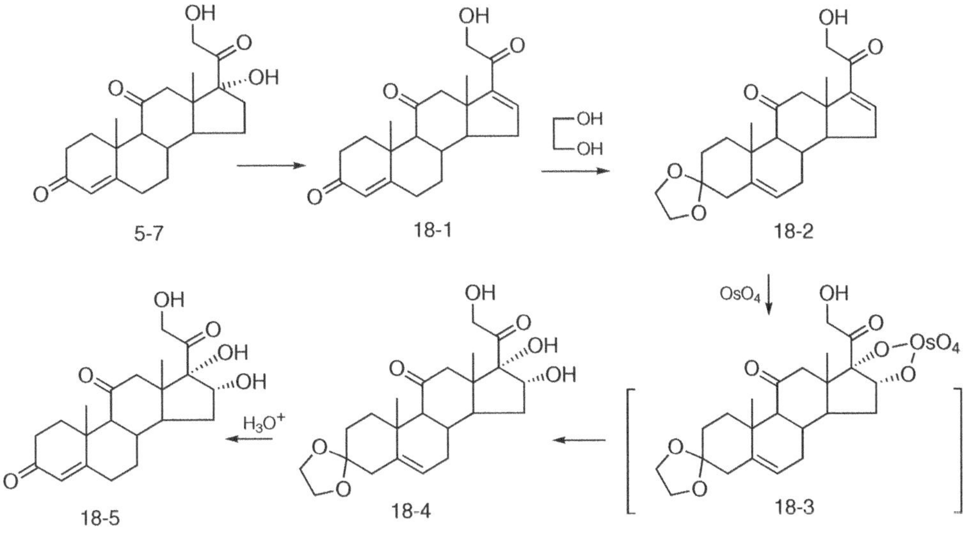

Scheme 7.18

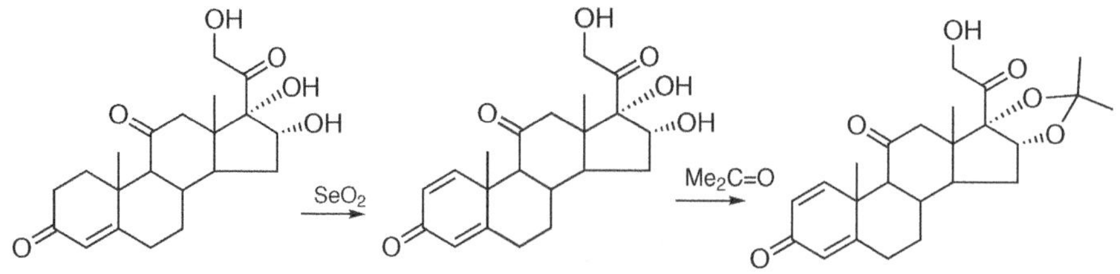

Scheme 7.19

group in ring A is then converted to the ethylene acetal **18-2**. This moves the olefin in ring A out of conjugation, and at the same time protects the C_3 ketone against conditions in the next reaction. Reaction of acetal **18-2** with osmium tetraoxide leads to attack of the reagent on the olefin in ring D. The resulting transient osmate derivative such as **18-3** is not isolated. This cyclic ether accounts for the formation of the regio- and stereochemistry of that transformation. This reagent is often used in catalytic amounts in the presence of another, more potent, oxidizing agent (see Scheme 7.4) for large-scale reactions. Work-up of the reaction mixture then affords the 16β,17β-glycol **18-4**. Hydrolysis with aqueous acetone cleaves the ethylene acetal to give 16β-hydroxycortisone (**18-5**).

Dehydrogenation with selenium dioxide gives the 1,4-dien-3-one (**19-1**). The great majority of such glycols are commonly used as their 16,17-cyclic acetals with some small ketone. In the case at hand, reaction with acetone gives the acetal **19-2**, that group often being called an acetonide.

7.3.5 Corticoids with Multiple Modifications

7.3.5.1 Starting from 6- or 16-Methylated Corticoids

Combining two changes, each of which increased potency of the resulting of the steroid, often leads to a corticoid that is more potent than either of the parent molecules. Construction of a compound substituted by a 9α-fluoro group and 16α-methyl group starts by conjugate addition of methylmagnesium bromide to the intermediate **12-1**. The organometallic, as expected, adds from the more open backside to give the 16α-methyl derivative **20-1** (Scheme 7.20). The 17-acetyl side chain is next converted to the requisite dihydroxyacetone by means of Gallagher chemistry (see Scheme 7.5). The alcohol group at C_3 is then oxidized to the corresponding ketone **20-3**. That function is protected by conversion to its bis ketal **20-5** by reaction with ethylene glycol under forcing conditions.

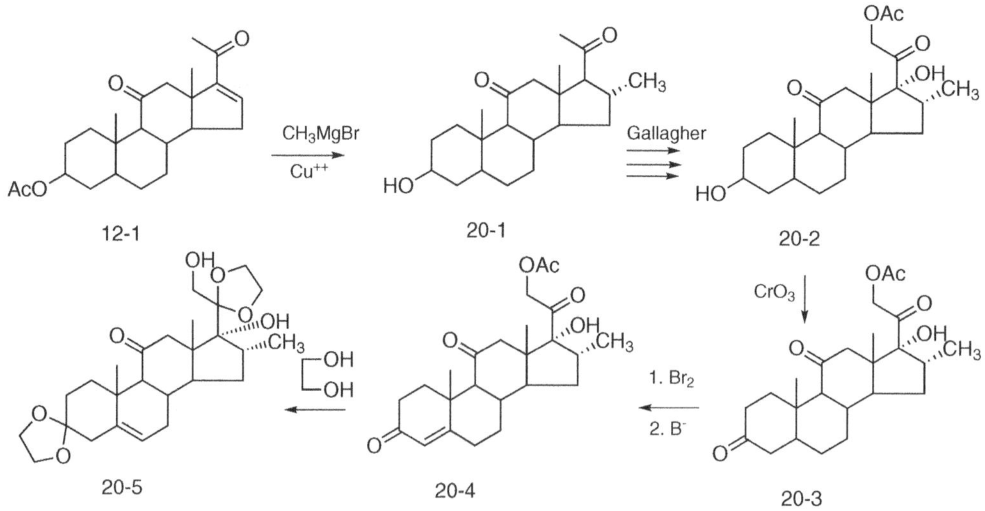

Scheme 7.20

Intermediate **20-5** is next subjected to the sequence used to incorporate a 9α-fluoro group (see Scheme 7.14) to afford **21-1** (Scheme 7.21). Hydrolysis of the acetal groups with mild acid restores the carbonyl groups at C_3 and C_{20} (**21-2**). Dehydrogenation of this last intermediate by means of selenium dioxide introduces an additional double bond in ring A; this product, dexamethasone (**21-3**), is yet another widely used corticoid.

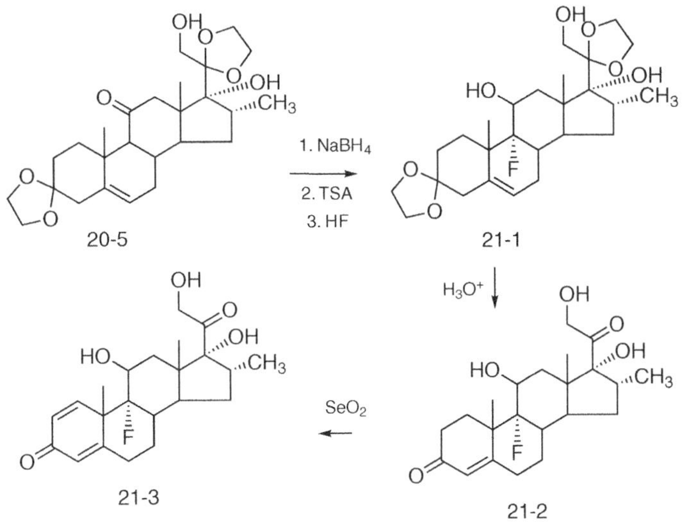

Pregnanes, Part 2: Corticosteroids **113**

Scheme 7.21

The epimer betamethasone (**22-2**), in which the configuration of the methyl group is reversed (16β-methyl), has much the same activity as the 16α-methyl isomer. Preparation of that corticoid relies on the same set of transforms used to synthesize **21-3**. The first several steps are intended to convert the ketone at C_{11} into a hydroxyl group. The carbonyl groups at C_3 and C_{20} in the advanced intermediate **12-6** are thus first converted to their ethylene ketals, the remaining ketone at C_{11} next being reduced with a hydride reagent (Scheme 7.22). Treatment with weak acid cleaves the two ketals (**22-1**). The fluorine atom is then introduced via the standard scheme outlined in Scheme 7.14 to finally yield betamethasone (**22-2**).

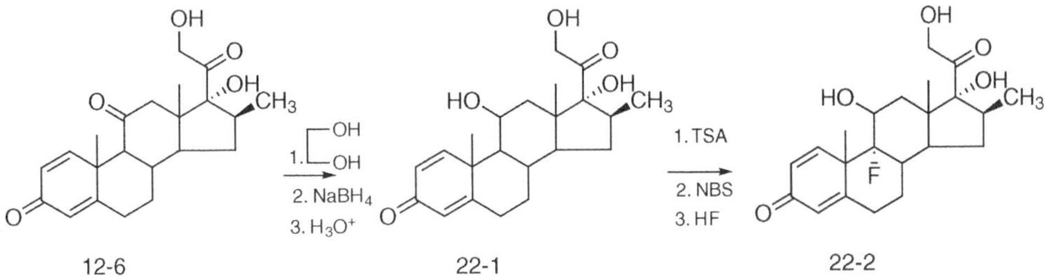

Scheme 7.22

The same general scheme for introducing the 9α-fluoro substituent is applicable to corticoids that feature a 6α-methyl group. 6α-Methylprednisolone (**9-4**) provides the starting material for the synthesis of fluorometholone (**23-5**) (Scheme 7.23). The standard sequence is again used to introduce the 9α-fluoro function (**9-4** → **23-1**). The only readily accessible hydroxyl at C_{21} is next converted to the mesylate **23-3** by acylation with methanesulfonyl chloride. Treatment of that intermediate with sodium iodide replaces the mesylate to afford the more readily reducible 21-iodo

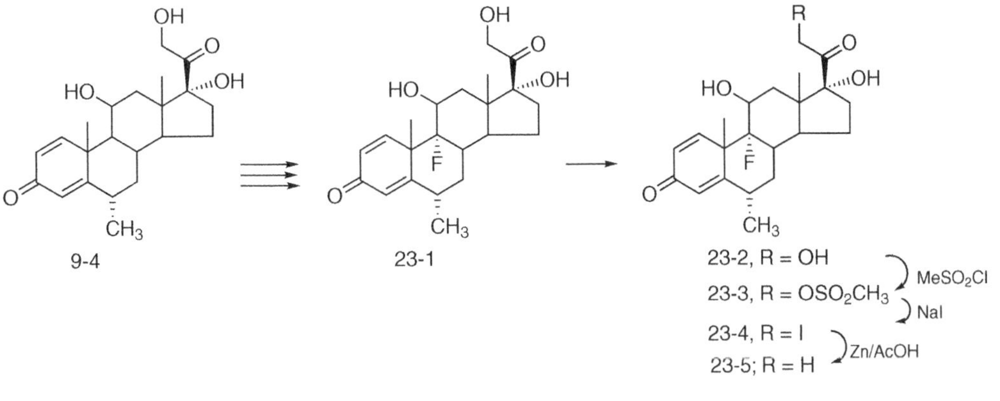

Scheme 7.23

derivative **23-4**. Zinc in acetic acid next removes the iodo substituent by reduction, giving **23-5**. The far less reactive 9α-fluoro function persists to this step unchanged.

Steroids with substituents on C_7 are actually a fairly rare species. Starting material for one such derivative comprises the 17-propionate **24-3**, prepared by a somewhat unusual sequence (Scheme 7.24). Reaction of the **24-1** with trimethyl orthopropionate can be rationalized by assuming that the first step involves exchange of the terminal hydroxyl of the side chain with one of the methoxyls in the orthoformate. The newly formed C–O bond holds the reagent in place so as to facilitate exchange of a second methoxyl on the reagent by the hydroxyl at C_{17} to afford the cyclic orthopropionate **24-2**. Mild acid causes that cyclic intermediate to open to the corresponding 17-propionate ester **24-3** The free hydroxyl group at the end of the side chain in this last product is next acylated by means of acetic anhydride; the conjugation of the C_3 carbonyl group is then extended by dehydrogenation with DDQ (**24-4**). Reaction of hydrogen chloride with this last product arguably proceeds by initial formation of a 1,6-adduct to the extended conjugated system to afford a transient intermediate such as **24-5**. Reversion of the 3-enol to a carbonyl group and shift into conjugation of the remaining olefin would afford the corticoid aclomethasone (**24-6**).

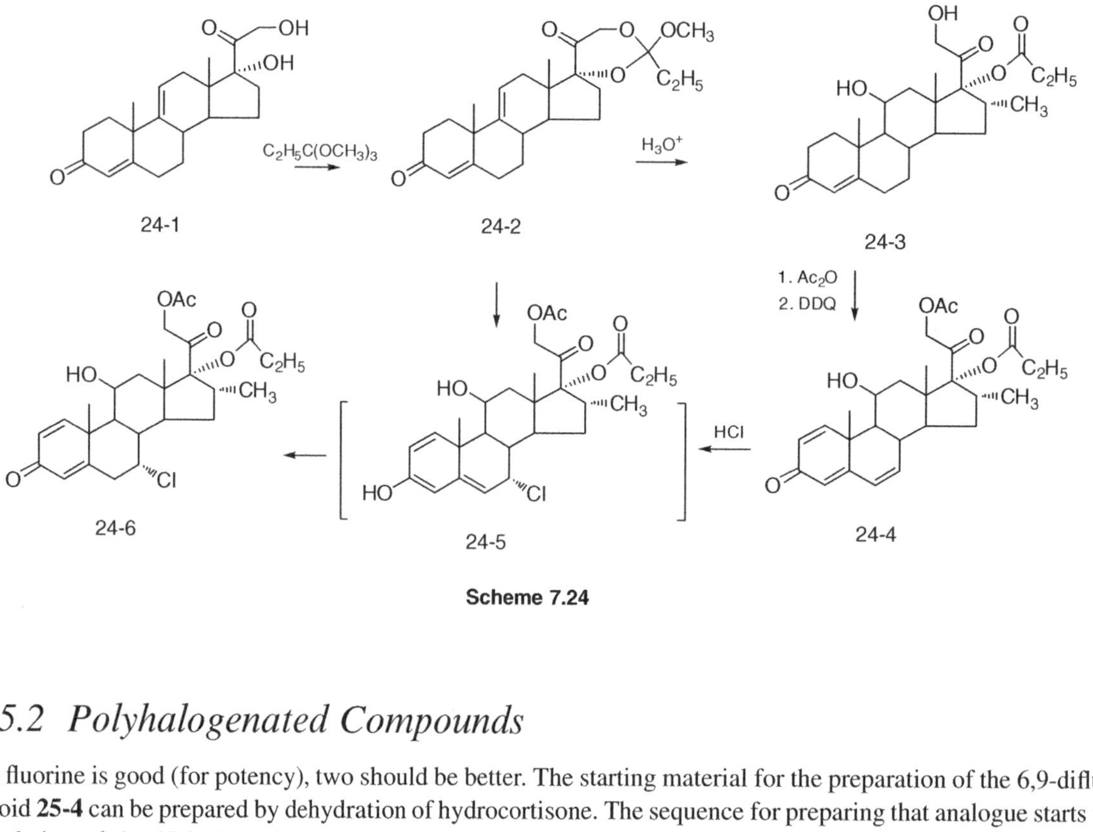

Scheme 7.24

7.3.5.2 Polyhalogenated Compounds

If one fluorine is good (for potency), two should be better. The starting material for the preparation of the 6,9-difluoro corticoid **25-4** can be prepared by dehydration of hydrocortisone. The sequence for preparing that analogue starts with the acylation of the 17-hydroxyl group by the same procedure as that used in Scheme 7.24 but substituting ethyl orthobutyrate for the orthopropionate (Scheme 7.25). The product from that transform, **25-1**, is next subjected to exhaustive acetylation; that step as expected, leads to acylation at the primary alcohol; the conjugated carbonyl group in this case reacts as well in this case, forming the enol acetate, **25-2** (see Scheme 7.16). Reaction with perchloryl fluoride then leads to the 6-fluoro derivative as a mixture of epimers at C_6. Strong acid converts the initial product to the favored equatorial 6α-fluoro derivative **25-3**. Application of the standard scheme for introducing the 9α-fluoro function then affords the corticoid, difluprednate (**25-4**).

Corticoids bearing halogen on C_2 constitute another rare species. Starting material for the rather lengthy synthesis of one of these corticoids is the fluohydrin **26-1**, obtainable from opening of a 5α,6α-epoxide derived from hydrocortisone. Treatment of this compound with bromine leads to the 2-monobromo derivative **26-2** (Scheme 7.26) (there is evidence that all bromination of saturated ring A steroids proceeds by initial formation of 2-bromides; the 2,4-dibromo derivatives used as intermediates in aromatizing ring A are thought to represent rearrangement products). The free hydroxyl at C_{11} is then converted to the mesylate **26-3** by means of methanesulfonyl chloride (**26-3**). Acetylation under forcing conditions give the 5,17-diacetate **26-4**. The very weak base sodium acetate then serves to eliminate the 5-acetate to give enone **26-5**. The fluorine epimerizes in the process to form the favored equatorial 6α-fluoro epimer.

Formation of a 2 bromo derivative by dehydrobromination requires, of course, the presence of an extra bromine at C_2. This halogen is provided by a second bromination step to afford the dibromo derivative **26-6**. Treatment of

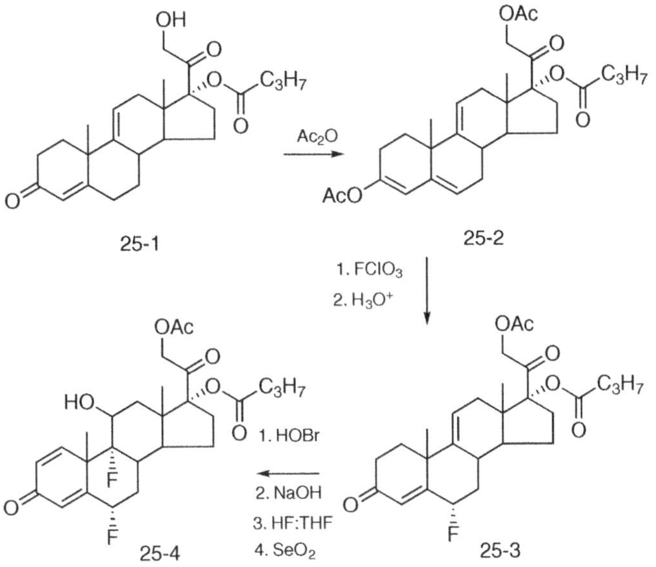

Pregnanes, Part 2: Corticosteroids 115

Scheme 7.25

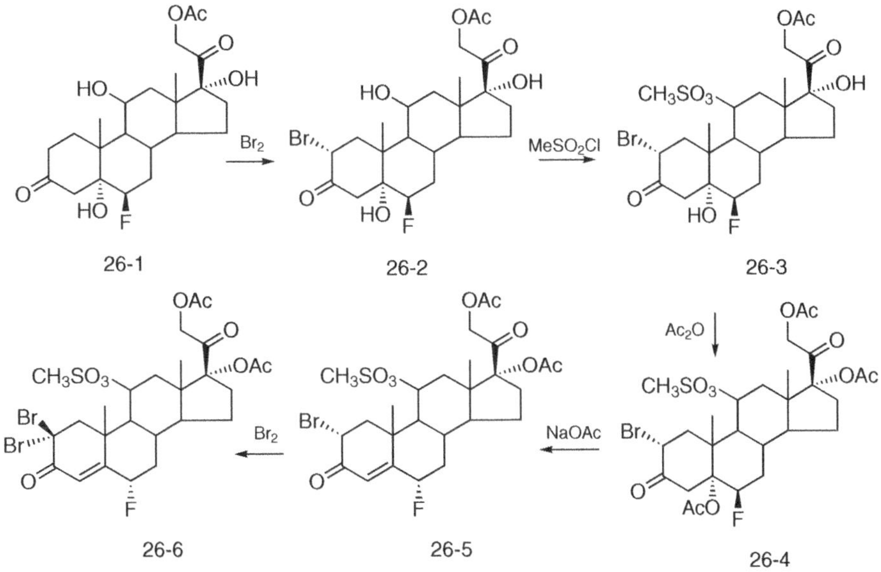

Scheme 7.26

this dibromo intermediate with lithium carbonate effects the desired dehydrobromination; the mesylate is removed in the process to form the requisite (9)11 double bond (**27-1**) (Scheme 7.27). The 9α-fluoro-β-hydroxy function is then generated by application of the by-now familiar sequence. This gives the corticoid haloprednone (**27-2**).

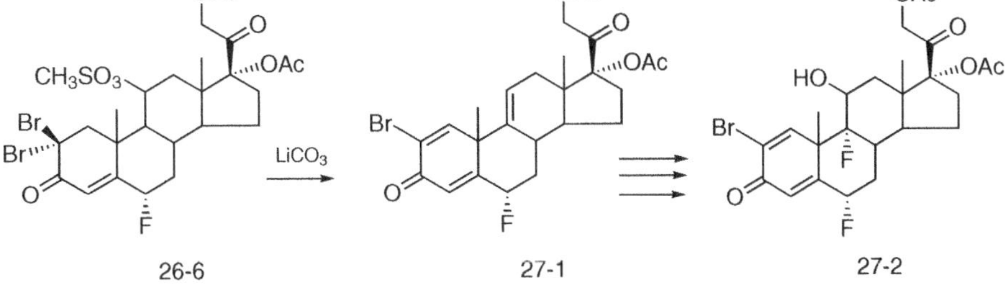

Scheme 7.27

Preparation of another trihalogenated corticoid starts with the advanced intermediate **28-1**, which is prepared by a route analogous to that depicted in Scheme 7.17. Reaction of that compound with chlorine leads to the formation of the 9α,11β-dichloro corticoid **28-2** (Scheme 7.28). The stereochemistry again reflects the intermediacy of a 9,11-chloronium species in the halogen addition step. Heating the product with selenium dioxide removes hydrogen from ring A to yield the 1,4-dien-4-one system. The product, **28-3**, is yet another anti-inflammatory agent.

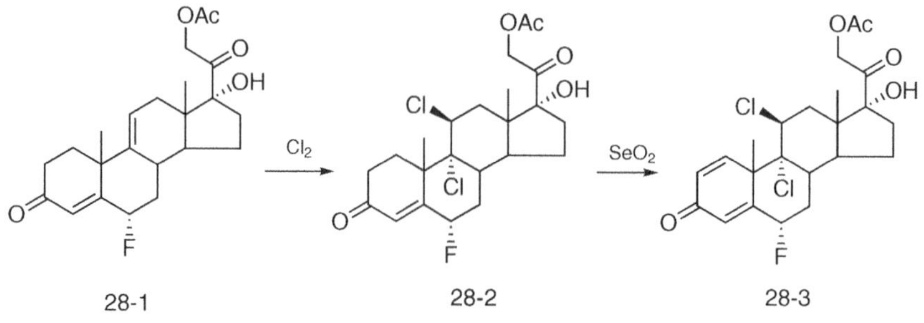

Scheme 7.28

7.3.5.3 Modified 16α,17α-Diols

The presence of a 16α-hydroxyl group, as noted earlier, enhances the potency of cortisone. Preparation of an analogue that includes a 9α-fluoro substituent in addition to the 16-hydroxyl starts with the 3,17-biscetal of hydrocortisone, **29-1** (Scheme 7.29). Treatment of that derivative with thionyl chloride in pyridine leads to dehydration of the hydroxyl groups at both C_9 and C_{17} to afford the 4,(9)11,16-triene **29-2**. Dilute acid then hydrolyzes the acetals to reveal the carbonyl groups at C_3 and C_{20}. Osmium tetraoxide or alternately potassium permanganate then selectively attacks the olefin in the five-membered ring from the more open backside, to afford the 16α,17α-diol (**29-4**). The isolated olefin at (9)11 is then taken on to the 9αfluoro-11β-hydroxy function by the standard scheme (**29-5**). Dehydrogenation by means of selenium dioxide inserts the extra olefin in ring A. The 16,17-*cis*-diol forms an acetal when treated with acetone (**29-6**). This product, triamcinolone acetonide, is a reasonably potent corticosteroid.

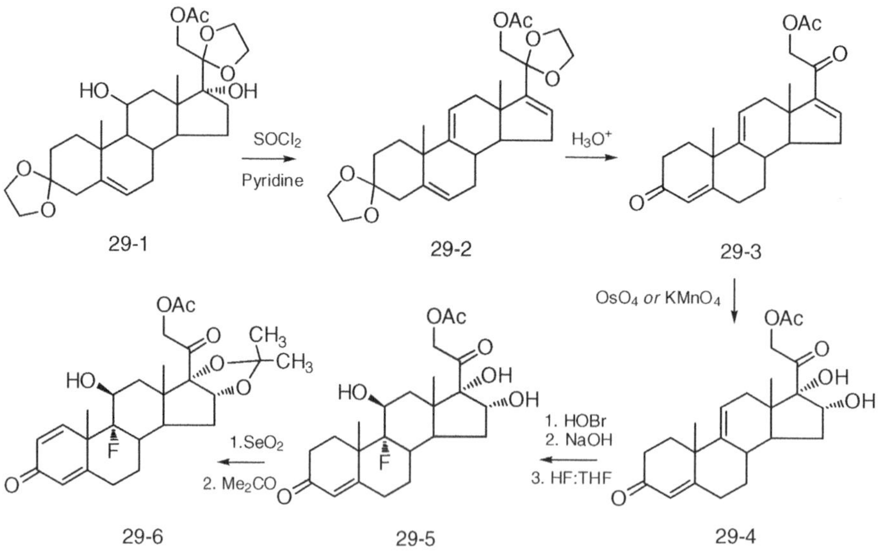

Scheme 7.29

An acetonide is used as a protecting group in the synthesis of a corticoid that incorporates no fewer than four halogen atoms. The sequence starts with the conversion of the terminal hydroxyl on the side chain at C_{21} to a better leaving group, a mesylate, by means of methylsulfonyl chloride (**30-2**) (Scheme 7.30). Treatment of that intermediate with lithium chloride displaces the mesylate to afford the 21-chloro analogue (**30-3**). Addition of chlorine then leads to the 9α,11β-dichloro derivative (**30-4**).

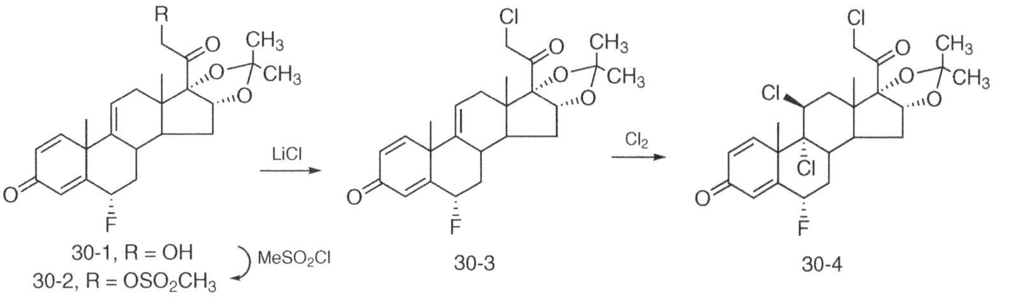

Scheme 7.30

Vilsmeyer formylation is not infrequently used for modifying alicyclic systems, although it is most frequently considered a reaction that applies to aromatic systems. The sequence for inserting an extra carbon atom by means of that reaction starts with conversion of the carbonyl group in **31-1** to its ketal by reaction with ethylene glycol; the double bond moves over to ring B, as is usually the case in this transformation (**31-2**) (Scheme 7.31). Treatment of that intermediate with N,N-dimethylformamide (DMF) and phosphorus oxychloride formylates the steroid at C_6, a position of high electron density. In the course of the reaction, the acetal opens and the thus-revealed hydroxyl is replaced by chlorine. The double bond at C_{5-6} stays in place because there is no longer the driving force – a ketone at C_3 – to cause it to shift back.

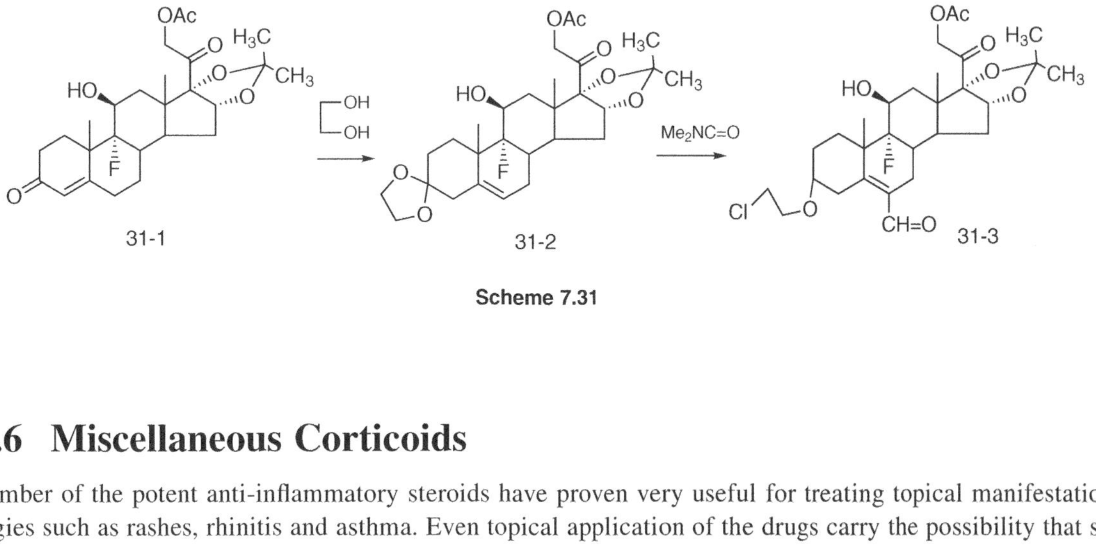

Scheme 7.31

7.3.6 Miscellaneous Corticoids

A number of the potent anti-inflammatory steroids have proven very useful for treating topical manifestation of allergies such as rashes, rhinitis and asthma. Even topical application of the drugs carry the possibility that some would be absorbed and find its way into the circulation, where it could cause the typical corticoid side-effects. Several compounds in both this and other unrelated therapeutic areas include functional groups that will be destroyed by serum enzymes, thus inactivating that portion of the topically applied compound that may have entered the circulation.

Replacement of side chain carbon C_{21} by thioester sulfur retains corticoid activity and at the same time provides a function that is destroyed by serum enzymes. The synthesis starts with the potent antiallergic agent flumethasone (**32-1**). Reaction of that steroid with periodic acid cleaves the terminal bond in the hydroxyacetone side chain to give the hydroxyl acid **32-2** (Scheme 7.32). Steric hindrance around the acid invoke the need for extra activation of that function. Reaction of **32-2** with diphenyl chlorophosphate thus provides the mixed anhydride **32-3**. This intermediate is not isolated but reacted *in situ* with N,N-thioformamido chloride. The transient new mixed anhydride, **32.4**, then undergoes an internal O to S rearrangement to give the acyl thioacid **32-5**. Saponification with sodium hydroxide affords the corresponding thioacid. Alkylation of that intermediate with fluoromethyl bromide then yields the fluoromethyl thioester fluticasone (**32-6**).

Reaction of the mixed anhydride **32-3** with methanethiol affords the corresponding thioester ticabesone (**33-1**), another topical antiallergenic drug (Scheme 7.33). It should be noted that the preparation another steroid for treating topical inflammation with sulfur substitution on ring D, **34-5**, was described in Chapter 5.

Yet another modification of the corticoid nucleus fuses a heterocyclic ring on to ring D. The starting 16β,17β-epoxide can be prepared by a scheme analogous to that used to prepare the 9α-fluoro-11β-hydroxy-modified steroids, namely

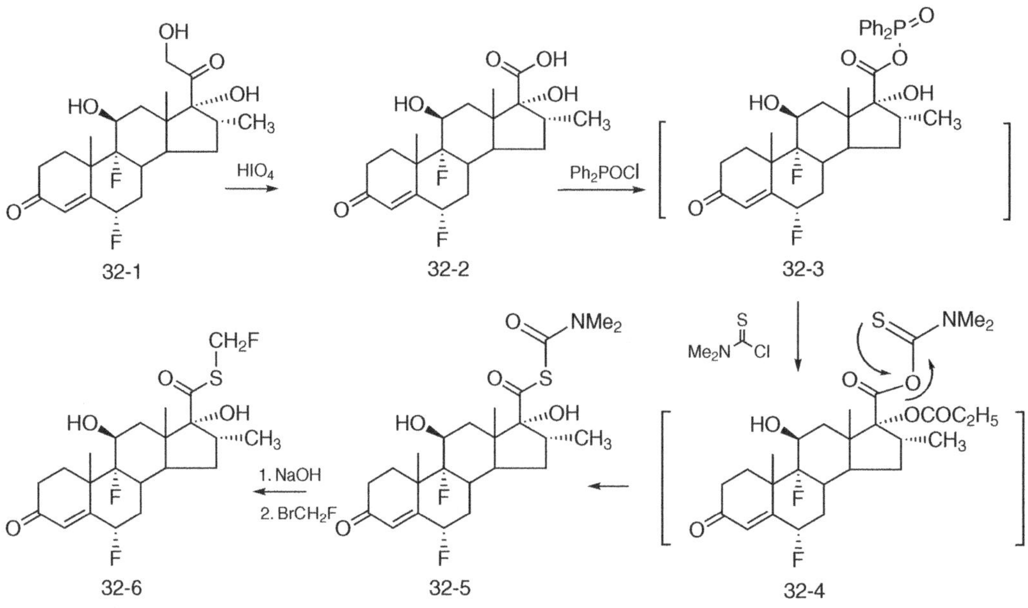

118 *Steroid Chemistry at a Glance*

Scheme 7.32

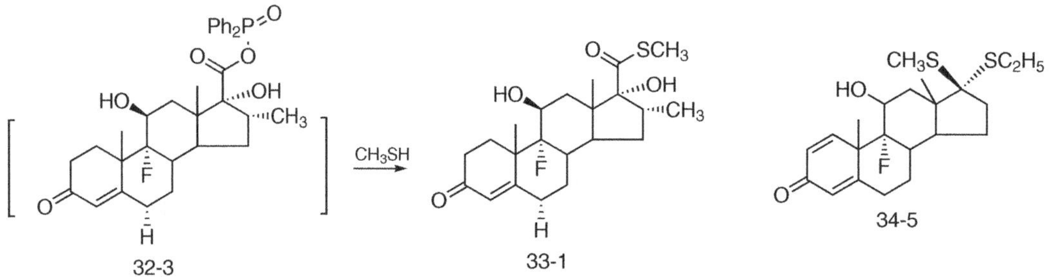

Scheme 7.33

successive reaction with HOBr followed by displacement of bromine by the adjacent alkoxide. Reaction of the epoxide by azide ion from its sodium salt opens the oxirane to form **34-2** (Scheme 7.34). The 16β-hydroxyl group apparently isomerizes under the reaction conditions to give the α-isomer. Reaction with acetic anhydride then yields the corresponding 16-acetate. Catalytic hydrogenation of the product reduces the azide to a primary amine (**34-5**). Treatment with acid causes the amine to attack the adjacent carboxyl group to form an oxazoline ring.

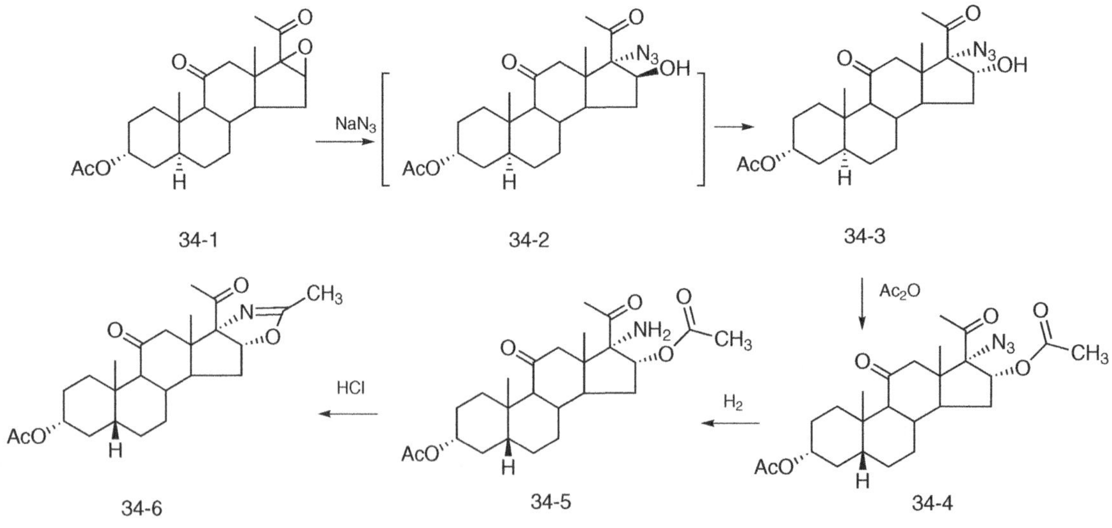

Scheme 7.34

The new fused ring also acts as a protecting group for the substituents on ring D. The chemistry for converting **34-6** to a corticoid starts with the saponification of the acetate at C_3 (Scheme 7.35). The newly freed hydroxyl group is then oxidized to the corresponding ketone. Bromination followed by dehydrobromination with a base such as lithium carbonate puts in place the 1,4-dien-3-one functional array (**35-2**). Several additional steps that include strategic protection-deprotection schemes build in the 9α-fluoro-11β-hydroxy moiety.

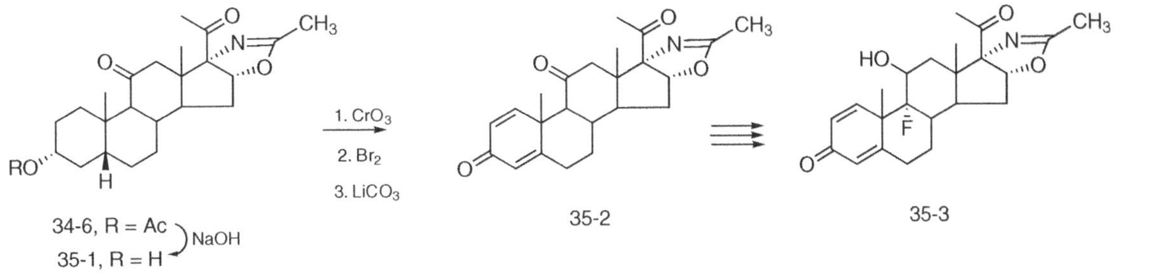

Scheme 7.35

7.4 Some Drugs Based on Corticoids

Corticosteroids, as noted at the beginning of this chapter, are best known for their ability to alleviate inflammation and to diminish allergic reactions. Corticoids that have been approved by regulatory agencies are, as a result, indicated largely for treating inflammation, allergic reactions or some combination of those afflictions.

The safety of topical formulations of some of the older corticoids has led to their approval for sale over the counter. Hydrocortisone acetate (**4-4**) formulated as a salve can, for example, be bought without prescription in drugstores, supermarkets and any other store that sells such products. Indications printed on a typical label start with the phrase 'For temporary relief of itching associated with minor inflammation and rashes due to ...' and then continue with a long list of possible causes of that syndrome.

Prednisone acetate (**6-2**) can in some respects be considered a workhorse among corticoid drugs. Although somewhat less potent than many of the more highly substituted molecules, the drug shows essentially the same qualitative range of activities as the latter. Prednisone acetate is available both as a formulation for oral administration and as an injectable solution. Patents covering the drug expired many years ago, which led to considerable competition from generic firms. Manufacturers of active drug substance have almost certainly improved the efficacy of the syntheses and may even possibly have developed new proprietary schemes for synthesizing prednisone. The end result is an Internet price for prednisone acetate of $11.95 for 30 tablets of a 20 mg dose, or about 40 cents per tablet.

Some medical conditions require the extended presence of a corticoid at the site of inflammation. One treatment for the pain from spinal stenosis or degenerative disk disease comprises epidural injection of a long-lasting corticoid. Depo Medrol®, a solution of 6α-prednisolone (**9-1**) in a special vehicle, is one of the drugs often used for that purpose. Once injected, the solution forms a depot at the site of administration; the drug slowly diffuses from that depot, leading to the sustained presence of the anti-inflammatory compound.

The potent corticoid triamcinolone (desacetyl-**29-5**) is also available in a range of formulations for the treatment of both allergy and inflammation. A triamcinolone salve, for example, is indicated for topical treatment of rashes and the like (unlike cortisone, this requires a physician's prescription). Treatment of asthma comprises an important indication for the drug. This use invokes the fact that the lungs originate embryologically from the same cell layer as skin. Triamcinolone is formulated as an inhalable powder for this purpose. Insuflation of the drug consequently constitutes topical administration

The very small amount of steroid that reaches the bloodstream from a topically administered drug can at least in theory cause some of the typical corticoid side-effects. This has occasioned research on corticoids that contain weak links that will lead to deactivation of the steroid by serum enzymes. The drug fluticasone (**32-6**) acts as a typical anti-inflammatory and antiallergic corticoid even though the side chain on ring D consists of a thioamide instead of a 2-hydroxyacetyl function. That thioamide provides the weak link that causes the drug to be destroyed by serum enzymes. As a result, fluticasone powder is used extensively in inhalers for treating asthma.

Beta-adrenergic agonists have been used for decades for treating asthma and related bronchial diseases by virtue of their relaxant activity on bronchioles. A product that consists of a combination of a beta-agonist with a corticoid would be expected to be more effective than either drug alone, since it would relieve bronchial spasm by two independent mechanisms. Several fixed-dose combinations of a potent corticoid with a beta-adrenergic agonist have been approved by the US FDA for treating asthma, bronchitis and chronic obstructive pulmonary disease (COPD). One of widely advertised drugs from this class, Adair® consists of a fixed combination of fluticasone and the beta-adrenergic blocker salmeterol.

8
Miscellaneous Steroids

8.1 Heterocyclic Steroids

8.1.1 Introduction

The journal and patent literature records a limited number of steroids in which one of the ring-carbon atoms has been replaced by oxygen and nitrogen. The pharmacological properties of many of these compounds are, with one notable exception, generally comparable to those of their carbocyclic models.

8.1.2 Steroids with a Heteroatom in Ring A

The electron density in furan rings is to some extent comparable to that of *O*-alkylated phenols. The starting material in the construction of an estrane in which the phenolic ring is replaced by the heterocycle (**1-1**) comprises a compound that has been used in steroid total syntheses (Scheme 8.1).

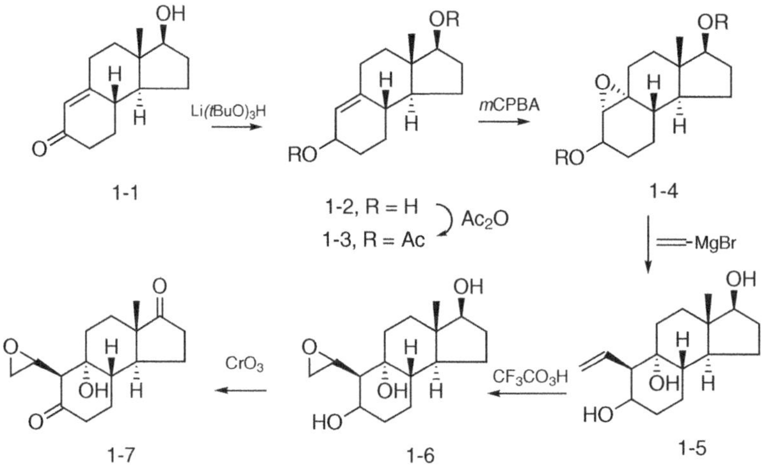

Scheme 8.1

Reduction of the conjugated carbonyl group in that compound with lithium tri-*tert*-butoxyaluminum hydride proceeds to the corresponding β-hydroxy isomer **1-2**. Acylation with acetic anhydride converts the alcohols to their acetates (**1-3**). The olefin is then oxidized to its epoxide **1-4** by means on 3-chloroperbenzoic acid. The stereochemistry for both the prior reduction and the peroxidation is determined by approach of the reagent, as in the case of steroids, from the sterically more open backside. Condensation of the epoxide with the Grignard reagent from vinyl bromide affords the isomer **1-5** from diaxial opening of the oxirane; the acetates are lost in the course of the reaction. A second epoxidation round, this time with the more potent oxidant trifluoromethylperacetic acid leads to the pendant epoxide **1-6**. Oxidation with chromium trioxide in acetone (Jones' reagent) affords the functional array required for rearrangement to a furan (**1-7**).

The boron trifluoride-catalyzed rearrangement of the epoxy ketone **1-7** to a fused furan can be rationalized by assuming that the Lewis acid complexes with the epoxide oxygen (**2-1**) (Scheme 8.2). Attack on the epoxide by oxygen from the enol form of the ketone will then close the five-membered ring. Dehydration of the resulting hydroxyl group forms the fused furan ring. The tertiary hydroxyl group also dehydrates under the reaction conditions to leave behind a double bond at C_9–C_{11} (steroid numbering). Reduction of the carbonyl group then leads to the 17β-hydroxy derivative

2-4. In much the same vein, condensation with lithium acetylide affords the corresponding 17α-ethynyl-17β-hydroxyl derivative. None of these target compounds showed much in the way of hormonal activity.

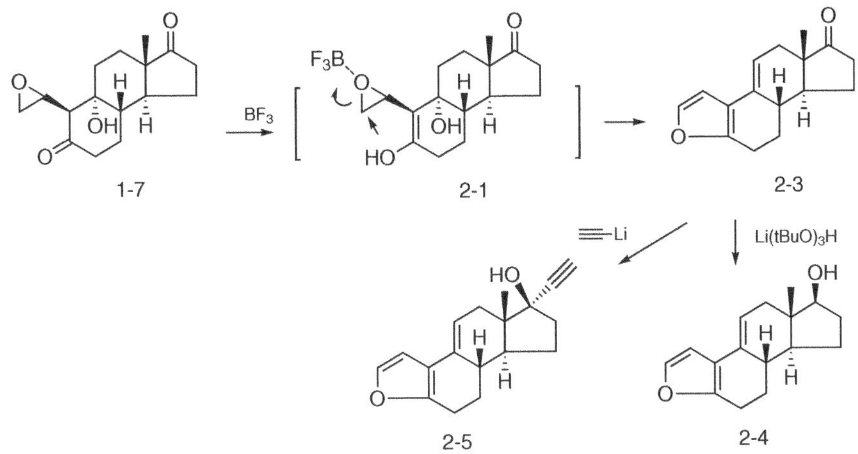

Scheme 8.2

Replacing C_2 in an androgen by oxygen while retaining the rest of the structure, by way of contrast, provides a potent androgenic–anabolic drug in spite the fact that the carbonyl group in ring A is now part of a potentially labile ester rather than a ketone. The sequence starts with the bromination of the fully saturated androstane **3-1** (Scheme 8.3). The resulting 2-bromo derivative is then dehydrobrominated with a base such as collidine to afford the conjugated derivative **3-2**. Treatment of that product with lead tetraacetate probably proceeds initially to hydroxylate both ends of the olefin to provide a 1,2-glycol such as **3-3**. Further oxidation then cleaves the bonds between hydroxylated carbon atoms in ring A. This results in expulsion of C_2 to leave behind a carboxaldehyde at C_{10} and an acetic acid moiety at C_5; this acid aldehyde exists in equilibrium with the hexalactol **3-5**. Reduction of the mixture of equilibrating species with sodium borohydride goes on to the observed lactone. This FDA-approved anabolic–androgenic agent has the USAN oxandrolone (**3-6**).

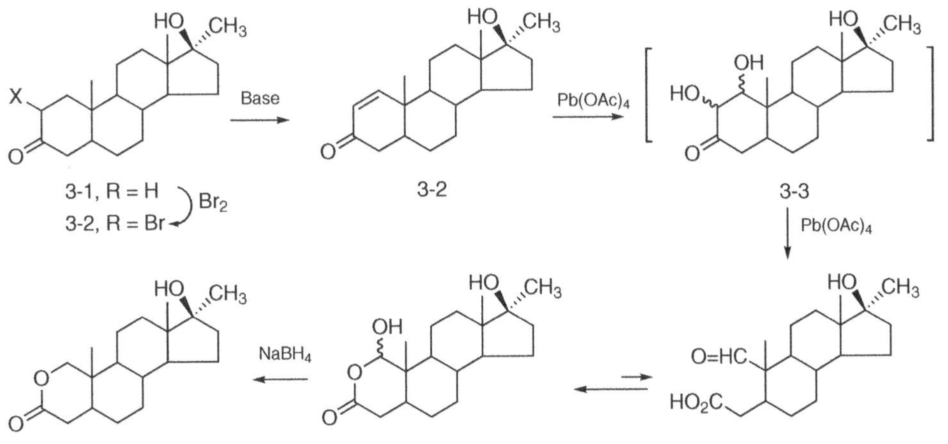

Scheme 8.3

Replacement of the carbon atom at C_4 by lactam-nitrogen also leads to compounds with pharmacological activity. The several FDA-licensed compounds in this small class act as androgen antagonists, in direct contrast to oxandrolone, which presumably acts on the same receptor.

The synthesis of the first of this group starts with the oxidation of the 17-acetyl side chain of progesterone (**4-1**) to a carboxylic acid by a haloform or some other related reaction (**4-2**) (Scheme 8.4). The hydroxy acid is then oxidized further by treatment with a mixture of potassium permanganate and periodic acid. This opens ring A and at the same time as above expels the superfluous carbon atom in that ring. This transformation can be rationalized by assuming that permanganate first hydroxylates the conjugated double bond (see above and Chapter 7, Scheme 7.29, for analogous reactions). Periodate then cleaves the diol to afford the observed keto acid **4-3**. This keto acid is next reduced

124 *Steroid Chemistry at a Glance*

catalytically in the presence of ammonia. The fact that the carboxylic acid at C_{17} comes through unchanged suggests that ammonia first converts the carbonyl group to an imine (**4-4**). This group is then hydrogenated to an amine; condensation with the carboxylic acid yields lactam **4-5**. The β stereochemistry of the amine follows from approach of hydrogen from the more open backside of the molecule. Reaction of the lactam with trimethylsilyl chloride then serves to convert the lactam to its trimethylsilyl ether **4-6**. There remains the task to add a double bond and an amide at C_{20}.

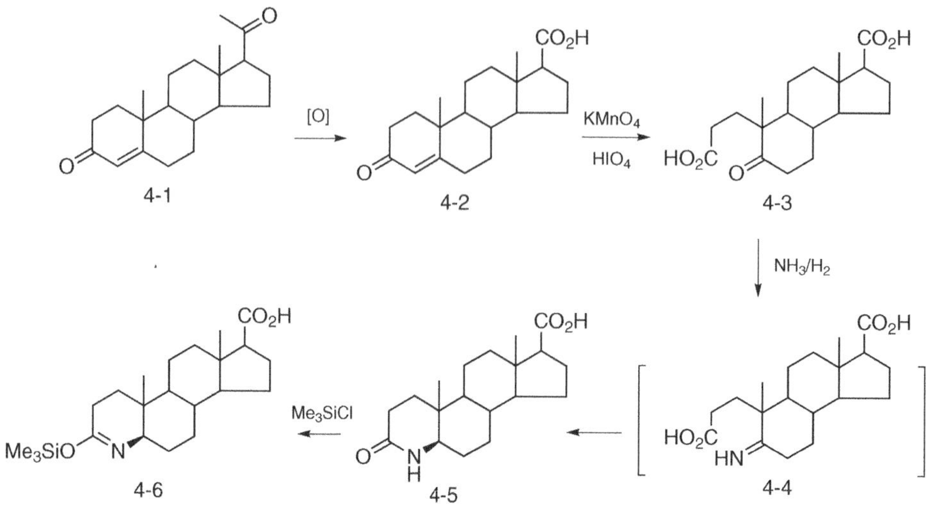

Scheme 8.4

Treatment of the silyl ether with DDQ introduces a second olefin in ring A; the silyl function in **5-1** is next removed by treatment with fluoride ion, specifically in this case with tetrabutylammonium fluoride (Scheme 8.5). Steric hindrance about the carboxylic acid invokes the need for extra activation of this group towards amide formation. The carboxyl group is therefore converted to its imidazolamide **5-3** by means of carbonyldiimidazole, and this imidazolamide is reacted with *tert*-butylamine. The resulting amide, finasteride (**5-4**), is a specific androgen antagonist that is used to treat conditions that will profit from diminishing stimulation by testosterone. This includes benign prostatic hypertrophy, male pattern hair loss and cancer of the prostate.

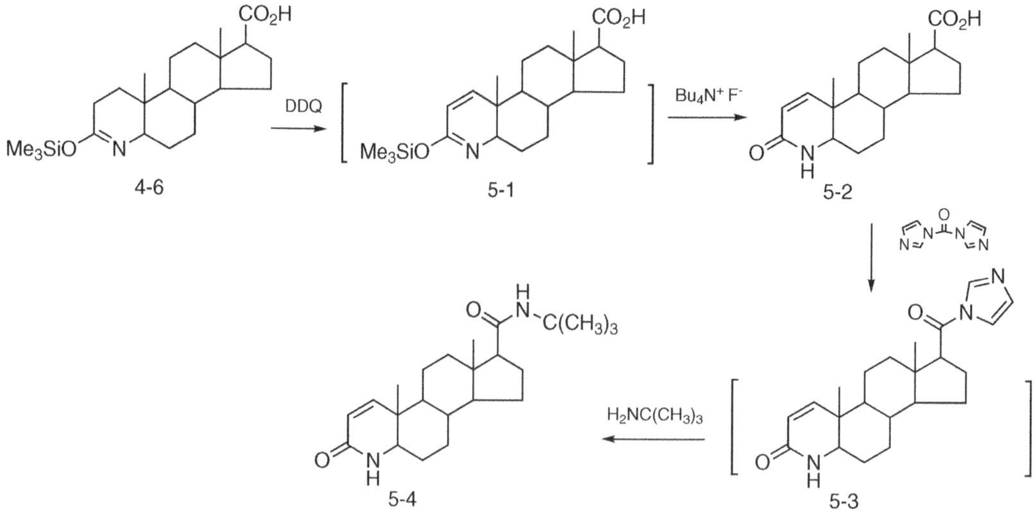

Scheme 8.5

8.1.3 Steroids with a Heteroatom in Ring B

Modifications of the Smith–Torgov scheme for total synthesis of estranes provides access to oxaestrane (**7-1**) and also its derivatives, **7-2** and **7-3**.

Preparation of the 6-oxa variant estrane starts with Lewis acid-catalyzed Friedel–Crafts cyclization of the arylpropionoxy acid **6-1** (Scheme 8.6). Condensation of that chromanone with the Grignard reagent vinylmagnesium bromide affords the carbinol **6-2**. As in the earlier work (see Chapter 3, Scheme 3.14), the crude product is next allowed to react with the surprisingly strong acid 2-methyl-1,3-diketocyclopentane in the presence of sodium carbonate to afford the adduct **6-4**. Treatment of this intermediate with ethanolic hydrogen chloride leads to a second cyclodehydration and formation of the 6-oxaestrane skeleton **6-5**. Catalytic hydrogenation selectively reduces the unsaturation in ring D (**6-6**).

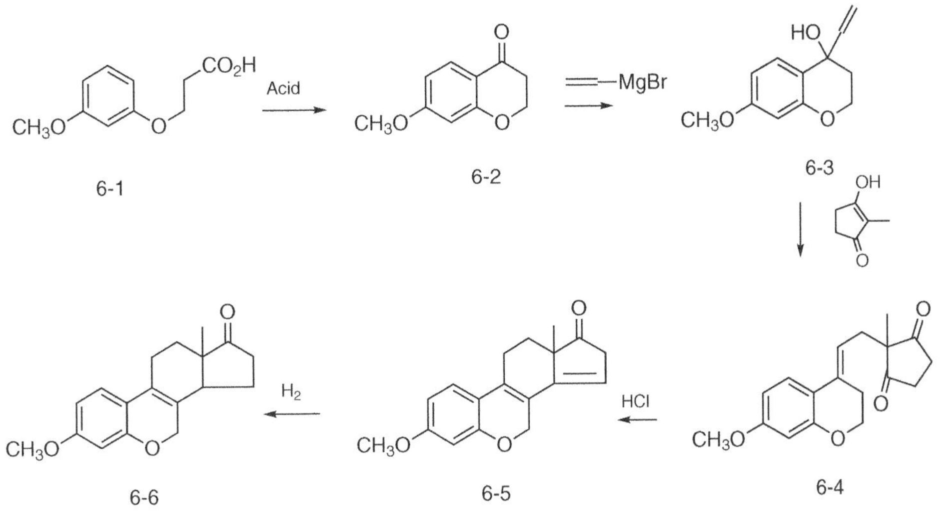

Scheme 8.6

The 8,9 double bond in the carbocyclic counterpart of **6-6** was reduced by means of lithium and an alcohol in liquid ammonia. Catalytic hydrogenation in the presence of acetic acid accomplishes the same step in the case at hand, affording **7-1** (Scheme 8.7). The observed *trans* rather than *cis* stereochemistry of the product is arguably due to the presence of acetic acid in the solvent in the hydrogenation step. The initial product likely features a *cis* B–C ring fusion The presence of the acid causes C_9 in the first-formed product to equilibrate so as to yield the favored *trans* 8–9 ring juncture. Sodium borohydride then reduces the carbonyl group to form the 17β-hydroxyl compound **7-2**. In the same vein, lithium acetylide adds to the carbonyl group to yield the 17α-ethynyl-17β-hydroxy derivative **7-3**.

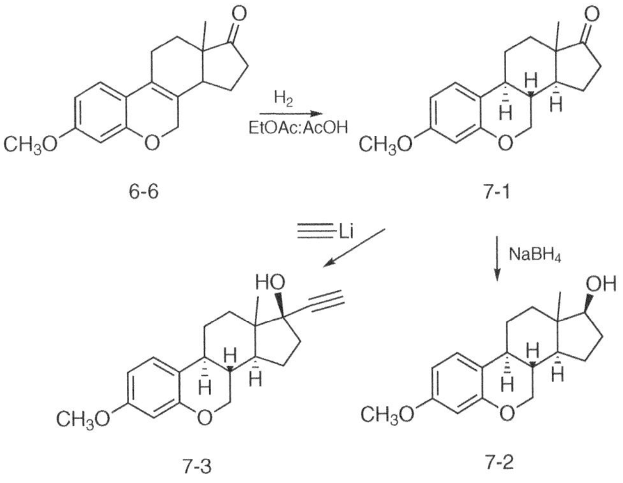

Scheme 8.7

Applying the same scheme to the phenylsulfonamido-protected partly reduced quinolone **8-1** proceeds much in the same manner as above, but only up to a point. Thus, Grignard addition of vinylmagnesium bromide proceeds to the adduct **8-2** (Scheme 8.8). Condensation of this intermediate with 1-methylcyclopenta-1,3-dione gives the direct azaestrane precursor **8-3**. This cyclizes on treatment with ethanolic hydrogen chloride to yield the azaestrane nucleus (**8-4**). Catalytic hydrogenation in this case also reduces the double bond in ring D. The carbonyl group is then converted to hydroxyl by means of sodium borohydride (**8-5**). The last double bond in the molecule proved resistant to reduction. Treatment under Birch conditions, however, removes the sulfonamide protecting group to afford the azaequilenin **8-6**. The formation of an extra double bond that aromatizes ring D in the product may represent a product from air oxidation

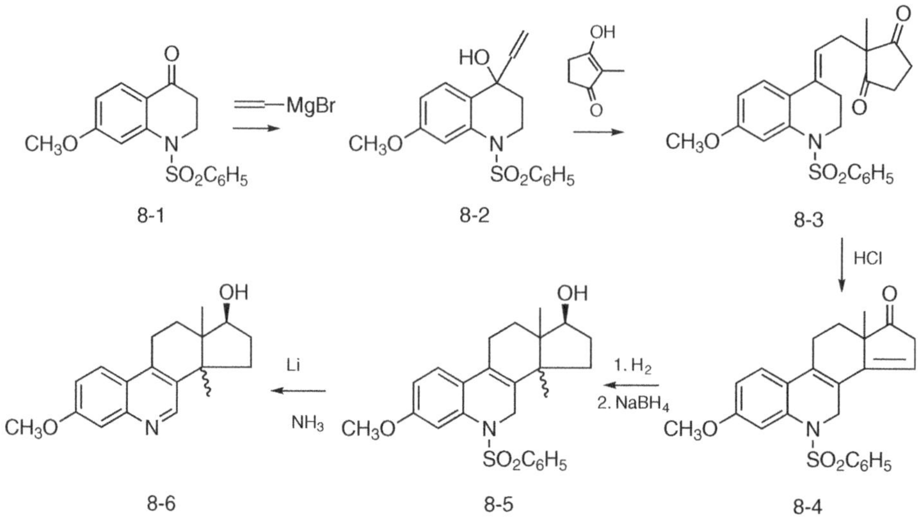

Scheme 8.8

8.1.4 Steroids with a Heteroatom in Ring C

The presence of oxygen in the form of a ketone or hydroxyl group at C_{11}, as noted in the previous chapter, has an important effect on the biological activity of pregnanes. The synthesis of an 11-oxaandrostane starts with a lengthy multistep degradation of the natural product hecogenin, the structure of which includes a carbonyl group at C_{12}.

Picking up the sequence at a later stage, ozonolysis of the unsaturated ketone **9-1** with oxidative work-up opens ring C to yield the keto acid **9-2** (Scheme 8.9). The remaining side chain is then cleaved, for example with periodate, to afford the corresponding 17-ketone **9-3**. The acetate ester is saponified and the resulting hydroxyl group oxidized with Jones' reagent (**9-4**). Heating the thus formed tricarbonyl intermediate with ethylene glycol leads to the bisketal; the ketone on ring B is apparently sufficiently hindered to resist acetal formation. The carboxyl group is next esterified with methanol. Treatment of the product with lithium aluminum hydride serves to reduce both the ester and the ketone on ring B. Treatment of the resulting dibasic alcohol with mild acid removes both ketals to afford **9-6**.

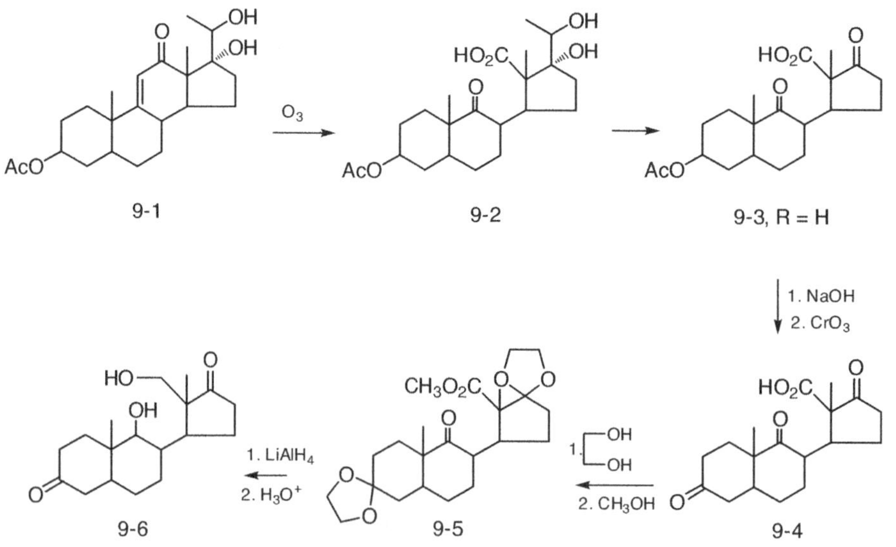

Scheme 8.9

Reaction of the diol **9-6** with *p*-toluenesulfonyl chloride and pyridine closes the ring to afford the cyclic ether **10-1** (Scheme 10.1). This transformation arguably starts by formation a tosylate of the primary alcohol; displacement by alkoxide at C_{10} (steroid numbering) forms ring C. Bromination of **10-1** with elemental bromine goes to a 2,4-dibromo derivative (not displayed). This intermediate is then dehydrobrominated to the customary 1,4-dien-3-one **10-2**. Hydrogenation over tristriphenylphosphinerhodium chloride selectively reduces the 1,2-double bond to afford the oxaandrostene **10-3**. The ketone at C_3 is then protected as its enol ether by means of ethyl orthoformate (**10-4**). Addition of lithium acetylide followed by hydrolysis finally gives the 11-oxaethisterone **10-5**, the 11-oxa analogue of ethisterone (see Chapter 5, Scheme 5.13).

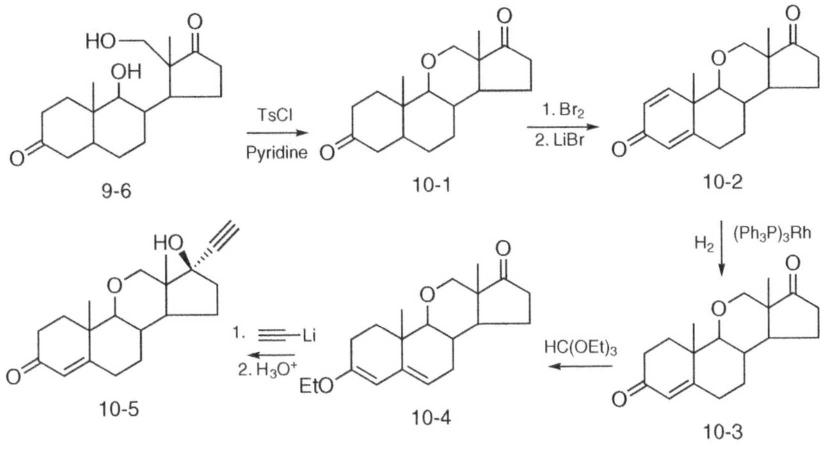

Scheme 8.10

A relatively recent scheme that involves an internal Diels–Alder reaction provides access to estrane-like compounds. These differ from their carbon-based prototypes by the presence of a hydroxyl instead of a methyl group at C_{13}. The synthesis starts with the acylation of the bissilyldiene **11-2** with chloroacetic anhydride (**11-1**) catalyzed by titanium tetrachloride (Scheme 8.11). This probably proceeds first to form a transient acylation product such as the ketone **11-3**. The presence of titanium tetrachloride in the reaction medium causes **11-3** to undergo internal acid-catalyzed aldol condensation then affords the cyclopentanol **11-4**. The treatment with base followed by hydrolysis gives diol **11-6**; alkylation of the primary hydroxyl with **11-4** leads to **11-8**. Heating the benzocyclobutyl intermediate **11-7** at a relatively mild 130 °C causes the strained four-membered ring to open (**12-1**) (Scheme 8.12). The diene from the opened cyclobutyl moiety and either one of two symmetrical symmetrically located pendant vinyl groups on the cyclopentyl fragment are now well located for an electrocyclic addition reaction. The resulting Diels–Alder reaction affords the steroid-like oxaestrane **12-2** in a 4:1 ratio with the 18α-alcohol. Oxidation of the vinyl fragment on ring D by means of a mixture of palladium acetate, benzoquinone and hypochlorite affords a mixture of the 17α-acetyl (**12-3**) and 17α-2-acetaldehydo (**12-4**) products.

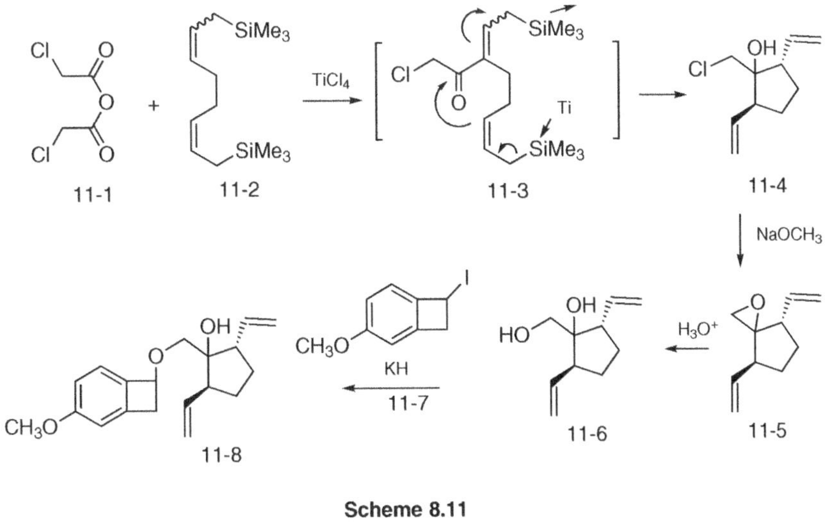

Scheme 8.11

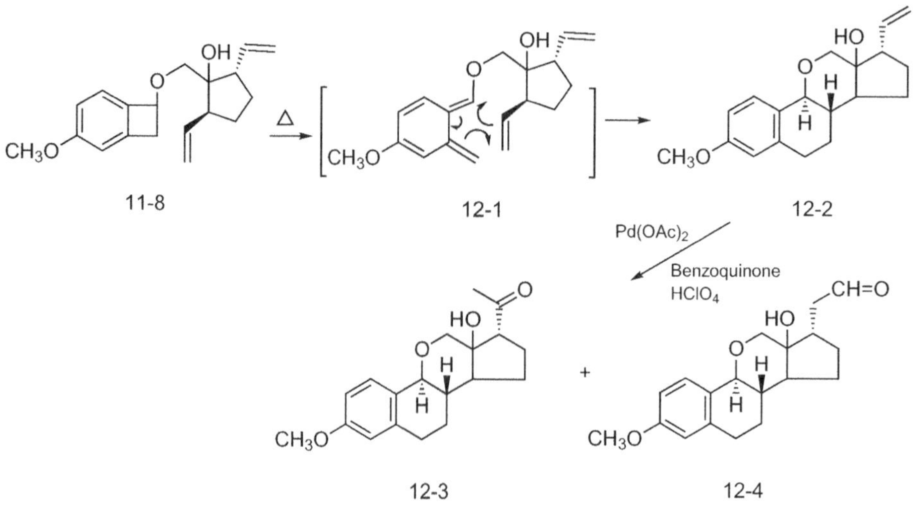

Scheme 8.12

8.1.5 Steroids with a Heteroatom in Ring D

Androgenic activity is lost when ring D of an androstane is converted to a lactone. The product, testolactone (**13-4**), has been used as an antineoplastic drug, but has been largely replaced by specific aromatase inhibitors. One preparation of this compound starts by dehydrogenation of testosterone with DDQ to form the 1,4-dien-3-one **13-2** (Scheme 8.13). The hydroxyl group at C_{17} is then oxidized by any of several methods such as using Jones' reagent. Bayer–Villiger reaction of the 3,17-dione with a peracid proceeds to form the lactone **13-4** exclusively.

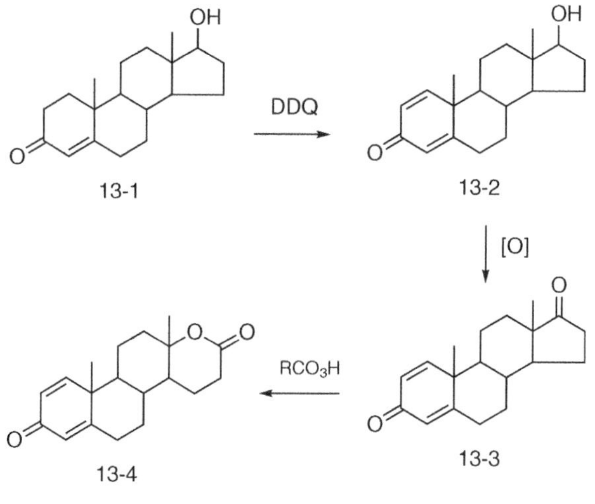

Scheme 8.13

A somewhat involved sequence for preparing 17-azaandrostanes begins with the conversion of the androstanolone **14-1** to its oxime, **14-2**, by means of hydroxylamine (Scheme 8.14). Treatment of that product with hydrogen chloride leads to Beckmann rearrangement of the oxime and consequent formation of the a derivative in which ring D has gone from a cyclopentanone to a ring-enlarged lactam (**14-3**). The regiochemistry of the product reflects the greater migratory aptitude of the quaternary center at C_{13}. The lactam nitrogen is next benzoylated by sequential reaction with sodium hydroxide followed by benzoyl chloride (**14-4**). The scheme for contracting ring D to a five-membered amine by extruding one carbon atom starts by formylation of the position adjacent to the carboxyl by reaction with sodium hydride followed by ethyl formate (**14-5**); the 3-acetate is hydrolyzed in the process. The crude product is next ozonized to yield the biscarbonyl derivative **14-6**. Reaction of this intermediate with peracetic acid then cleaves the bond that connects the two carbonyl groups; the newly generated *N*-formyl moiety is lost in the process to afford the acid amide **14-7**. Treatment of this intermediate with lithium aluminum hydride reduces the acid to a carbinol and the benzamide to an *N*-benzylamine, **14-8**.

Reconstruction of five-membered ring D is accomplished by reaction of the amino alcohol **14-8** with thionyl chloride. The reagent likely first interacts with the carbinol either to replace oxygen by chlorine or alternatively to form a

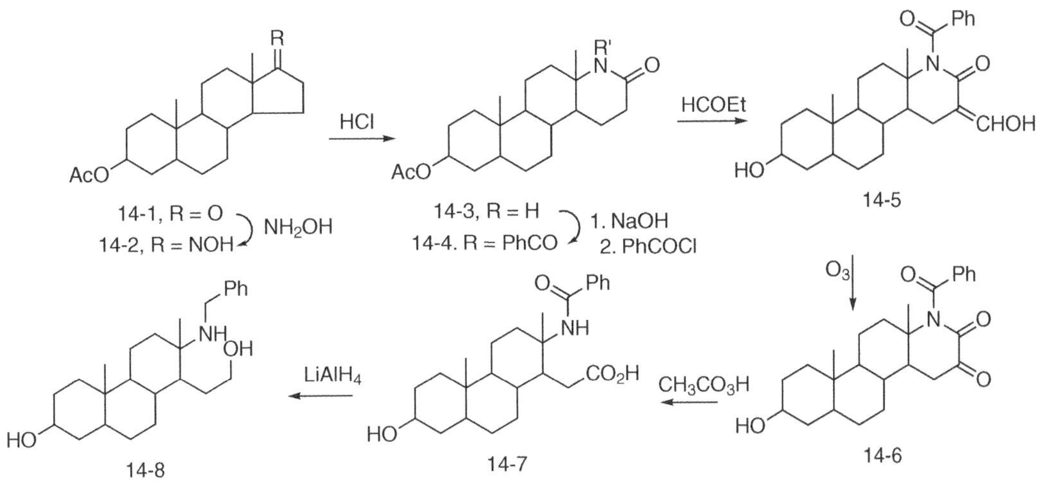

Scheme 8.14

sulfoxide-based leaving group. Displacement of halogen (or the sulfoxide) by basic nitrogen on C_{13} then closes the ring (**15-1**) (Scheme 8.15). The benzyl group on nitrogen is removed by catalytic hydrogenation, then acylation by means of acetic anhydride affords **15-2**. This can be transformed to 17-azaprogesterone by installing the 4-en-3-one function in ring A by one of the several schemes outlined in Chapters 5 and 6.

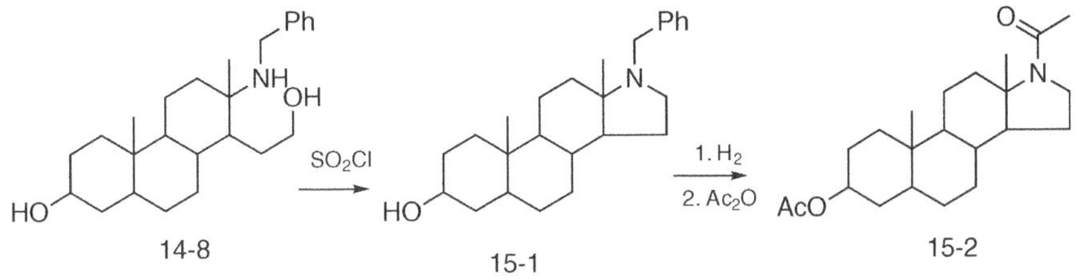

Scheme 8.15

8.2 Cardenolides

The drug digitoxin, the principal entity among the cardiac glycosides present in foxglove leaves, has a long history as a folkloric treatment for heart disease; powdered leaves from the plant were at one time used for just this purpose. The active substance, digitoxin, was more recently isolated from extracts of foxglove leaves. The drug increases the force of cardiac contraction (inotropic activity) and as a result has often been used to bolster a failing heart. The drug also found some use in treating arrhythmias and cardiac fibrillation. The effective dose of digoxin and structurally related agents is, however, uncomfortably close to that which is toxic; doses are therefore adjusted by tracking drug blood levels. These measurements arguably gave rise to the science of pharmacokinetics. The close relative digoxin (Scheme 8.16), also obtained from plants, acts more quickly and has largely displaced digitoxin in the clinic. Each member of this class of natural products consists of a steroid-like moiety linked to sugars by an acetal-like linkage. The sugar-free steroidal moieties or aglycones in this series are known collectively as cardenolides. The trivial names for these, aglycones, as a rule carry the suffix 'genin'. The recent availability of more specific and less toxic agents for treating heart disease has placed a damper on work intended to modify cardenolides in the search for better tolerated agents. The structures of these molecules do, however, provide a challenging goal for organic synthetic chemists. Virtually all of the synthetic work in this area has been devoted to modification of the aglycones.

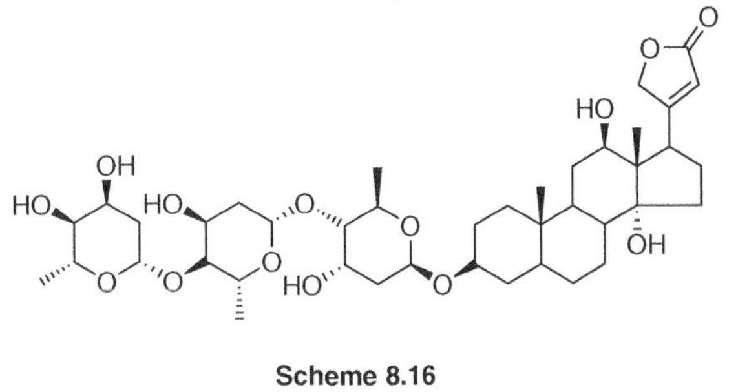

Scheme 8.16

8.2.1 Actodigin Aglycone

Limited structure–activity studies have demonstrated that the minimal substituents for cardiac activity include a β-hydroxyl group at C_3, an α-hydroxyl group at C_{14} and a butenolide at C_{17}. Inotropic activity is retained when the attachment of the butenolide to the steroidal nucleus is reversed from that in the natural product. Synthesis of an analogue in which attachment of the butenolide is reversed starts by hydrolysis of digitoxin to remove the sugar to afford the aglycone **17-1** (Scheme 8.17). Reduction of that intermediate with diisobutylaluminum hydride opens the furan ring to yield the 1,4-glycol **17-2**. Treatment of **17-2** with sulfur trioxide–pyridine complex selectively oxidizes the allylic hydroxyl group on the side chain to afford the transient hydroxyaldehyde **17-3**. That spontaneously goes on to form an internal acetal between the hydroxyl function and the aldehyde; loss of water then gives the observed product, furan **17-4**. (the furan ring is now displayed in rotated form)

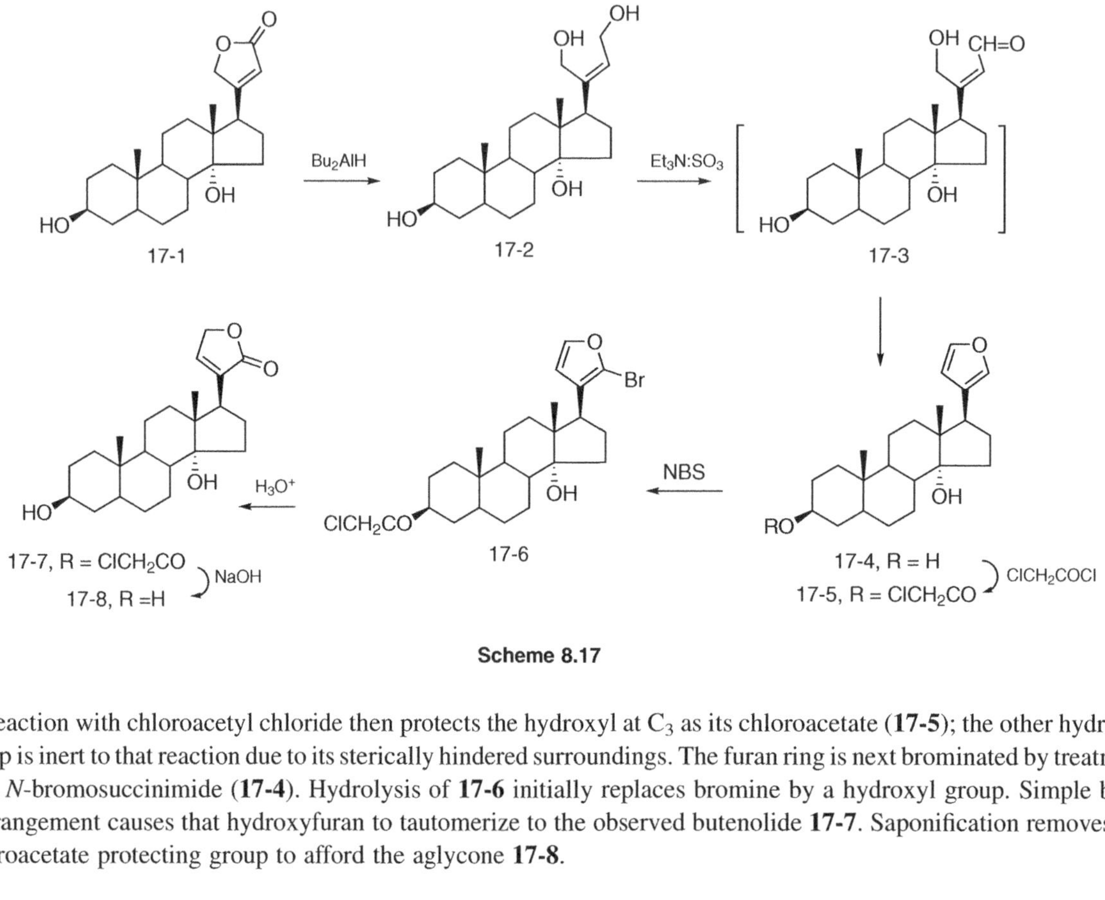

Scheme 8.17

Reaction with chloroacetyl chloride then protects the hydroxyl at C_3 as its chloroacetate (**17-5**); the other hydroxyl group is inert to that reaction due to its sterically hindered surroundings. The furan ring is next brominated by treatment with *N*-bromosuccinimide (**17-4**). Hydrolysis of **17-6** initially replaces bromine by a hydroxyl group. Simple bond rearrangement causes that hydroxyfuran to tautomerize to the observed butenolide **17-7**. Saponification removes the chloroacetate protecting group to afford the aglycone **17-8**.

8.2.2 Synthesis from a Bile Acid

A number of naturally occurring cardenolides such as that from digoxin incorporate a hydroxyl group at C_{12}, a substituent also present in bile acids. The somewhat lengthy synthesis of the aglycone present in digoxin invokes the use of three steps that involve irradiation. The sequence starts with conversion of deoxycholic acid **18-1** to its imidazolide by reaction with carbonyldiimidazole (Scheme 8.18). Two atoms are next expelled from the side chain by irradiating the compound in the presence of oxygen. The transformation can be rationalized by assuming that that it involves first photo-oxidation of the carbon next to the imidazolide **18-3**; that intermediate would then collapse to the observed product **18-4**. Mitsonobu reaction of that intermediate with acetic acid displaces the hydroxyl group at C_3 with inversion of stereochemistry to give the 3β-alcohol. A second photo-oxidation step, this one in the presence of a sensitizer, gives the allylic hydroxyl by way of attack by singlet oxygen (**18-6**).

Oxidation of the allylic hydroxyl in **18-6** by manganese dioxide proceeds in a straightforward fashion to aldehyde **19-1** (Scheme 8.19). The additional carbon atom required for the furan is added by Wittig condensation with the phosphorane from triphenylmethoxymethylphosphonium chloride and base. The product, **19-2**, is then subjected to yet another photo-oxygenation reaction. Singlet oxygen in this case adds across the diene to form the cycloadduct, peroxide **19-3**. The 1,2-dioxane **19-3** can be viewed as a latent hydroxy ester. Treatment of that compound with base causes the peroxide to break and the newly revealed functions to cyclize to give the cardenolide **19-4**. There remains the task of

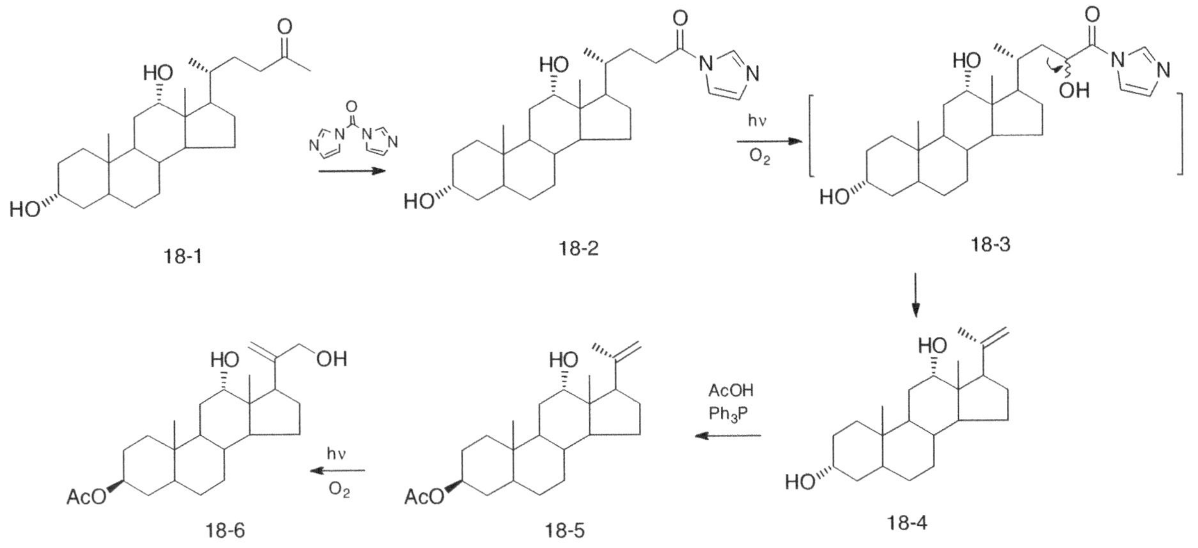

Scheme 8.18

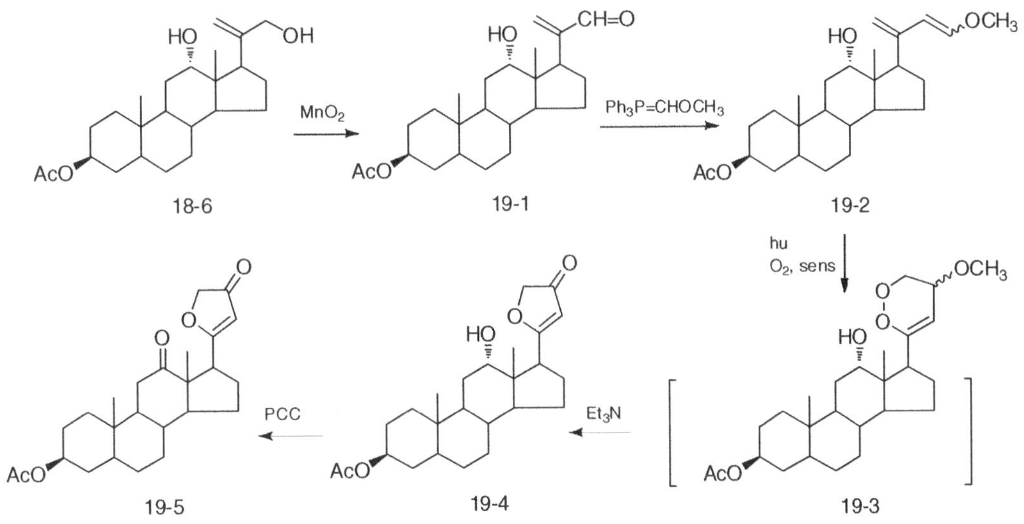

Scheme 8.19

adding the requisite hydroxyl group at C_{14}. The scheme for achieving this starts by oxidation of the hydroxyl group in ring C with pyridinium chlorochromate (**19-6**).

By way of review, the Prins reaction (Scheme 8.20) comprises addition of a carbonyl group to an olefin followed by capture of a nucleophile by the other end of the double bond (**20-1**). A side reaction consists of addition of carbonyl oxygen (**20-2**). Yet another photolysis, this one on **19-5**, severs the bond between the carbonyl and the adjacent quaternary carbon atom to the unsaturated aldehyde **21-1** (Scheme 8.21); this is admixed with the product of **21-2** from addition via oxygen. Treatment of the mixture of these products with mild acid leads to intramolecular Prins condensation of the aldehyde in **21-1** with the C_{13}–C_{14} double bond. Capture of a nucleophile, in this case water, from the less hindered face installs the critical 14α-hydroxyl function. The *O*-addition product **21-2** presumably cycles back to **21-1**, which the goes on to product. The 12β-hydroxy product predominates over its 12α-isomer in a 2:1 ratio.

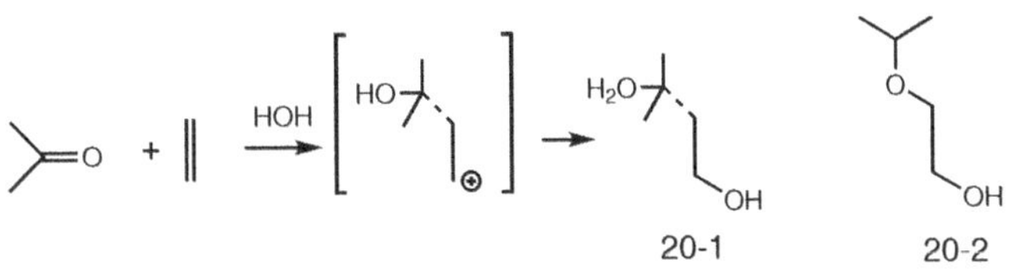

Scheme 8.20

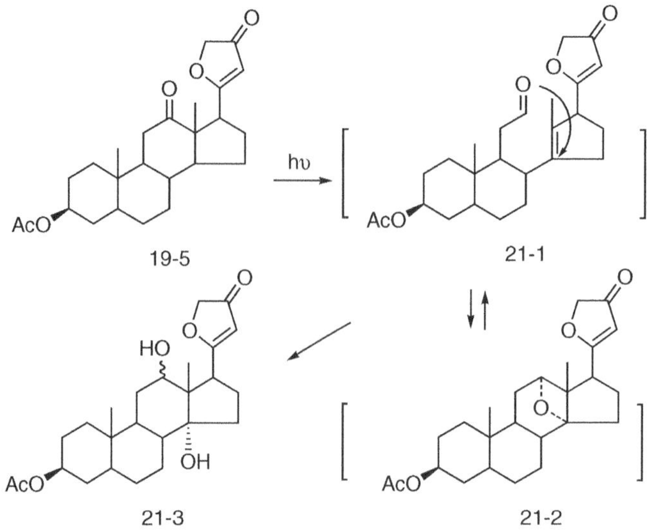

Scheme 8.21

8.3 Compounds Related to Cholesterol

Many strategies have been followed to control levels of serum cholesterol. A compound that incorporates the carbon skeleton of cholesterol itself, colestolone, is thought to inhibit cholesterol synthesis in a late step in its biosynthesis by acting as a product feedback inhibitor. The synthesis starts with bromination of the allylic C_7 position in cholesterol benzoate **22-1** by means of *N*-bromosuccinimide (**22-2**) (Scheme 8.22). Dehydrobromination, for example with collidine, leads to the endocyclic 5,7-diene **22-3**. In the presence of strong acid, the bonds migrate to form the transoid 8,9–14,15 diene **22-4** in very modest yield. The driving force in this reaction may come from the stability of the all-*trans* diene. Oxidation with chromium trioxide interestingly proceeds on both carbon atoms at the extremities of the diene system; the product thus features a hydroxyl group at 9α and a ketone at C_{15} (**22-5**). Treatment with zinc in acetic acid removes the allylic hydroxyl at position 9 by reduction (**22-6**). Saponification then cleaves the benzoate to afford colestolone (**22-7**).

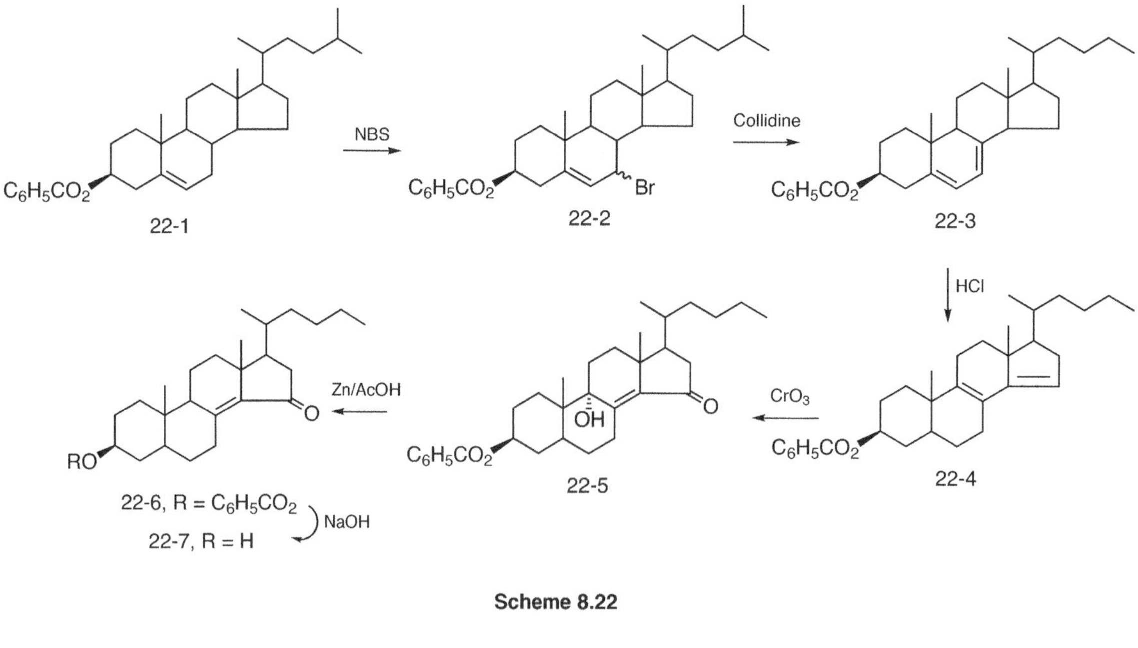

Scheme 8.22

Vitamin D plays a very central role in maintaining the integrity of bone structure. The vitamin actually comprises a set of closely related derivatives of cholesterol that incorporate a 5,7-diene in ring B but differ in the pattern of hydroxyl groups in ring A and on the side chain. The vitamins are not active *per se* but comprise precursors that require activation by ultraviolet irradiation. In a typical example, the action of light on so-called 7-dehydrocholesterol (**23-1**) cleaves the bond between rings A and C in a reverse cycloaddition reaction (Scheme 8.23). The initial product, **23-2**, then undergoes rotation about the former 6,7 bond to lead to a transoid orientation of the exocyclic double bonds. This product cholecalciferol (**23-3**) is one of the D vitamins. The agent acts as a direct regulator of calcium metabolism. The compound is now included as an ingredient in one of the phosphorus-based 'endronate' drugs for treating osteoporosis.

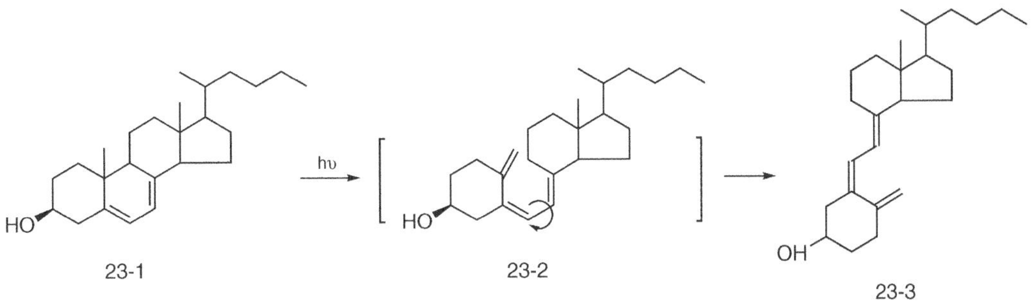

Scheme 8.23

Addition of hydroxyl groups to the side chain are known to increase potency in the vitamin D series. The starting material **24-1** can in principle be obtained from pregnenolone by extension of the side chain by a scheme that starts by addition to the carbonyl group at position 20. The diene in ring C can be elaborated by a scheme similar to that used to go from **22-1** to **22-3**. Reaction of the aldehyde in **24-1** with isopropenylmagnesium bromide will the lead to carbinol **24-2** (Scheme 8.24). The double bond is then oxidized to its epoxide **24-3**. Reduction with lithium aluminum hydride then opens the epoxide to afford the carbinol **24-4**; removal of the silyl protecting group with fluoride ion affords **25-5**. Irradiation of this last intermediate opens ring B; thermal isomerization then gives the transoid form secalciferol (**24-6**).

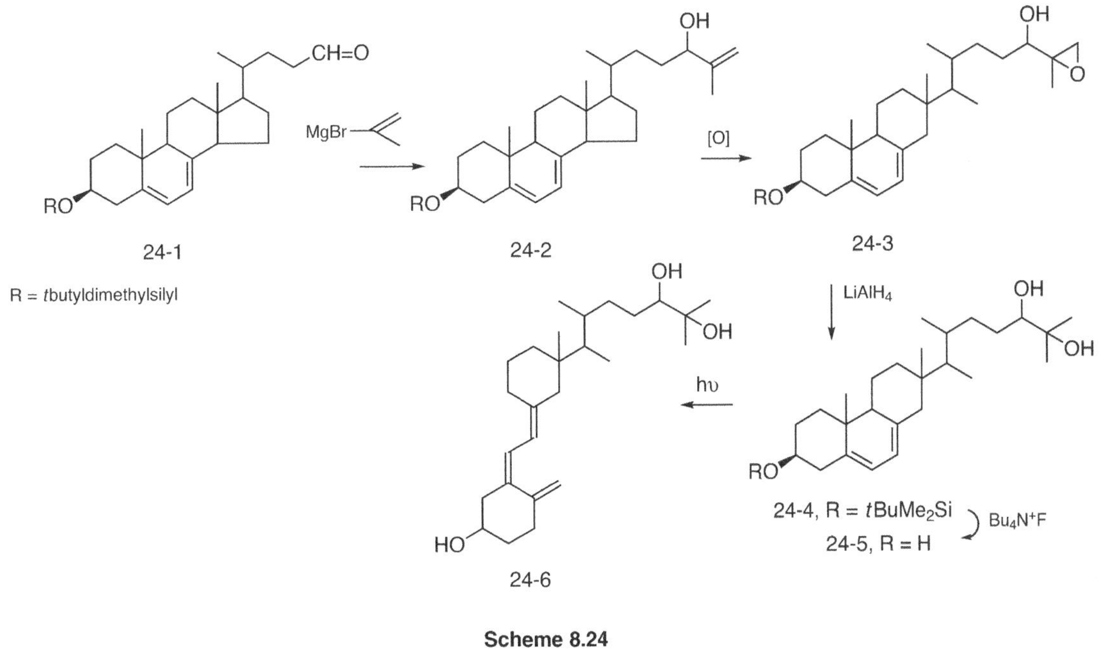

Scheme 8.24

Compounds related to vitamin D are also involved in regulation of skin growth. One of the D vitamins, calcitriol, has been used to treat psoriasis and acne. A recent semi-synthetic vitamin D congener has shown an improved therapeutic index over the natural product. The synthetic sequence to this analogue hinges on selective scission of the isolated double bond in the side chain at C_{17}. The key step thus involves inactivation of the conjugated diene centered on ring A. This is accomplished by formation of a Diels–Alder-like adduct between sulfur dioxide and the starting material **25-1**, in which the hydroxyl groups are protected as their diisopropylsilyl ethers (Scheme 8.25). Ozonolysis of the adduct **25-2** followed by work-up of the ozonide gives the chain-shortened aldehyde. Heating the product restores the diene by reversing the original cycloaddition reaction (**25-3**). The carbonyl group is then reduced to the alcohol by means of borohydride and the resulting alcohol is converted to its mesylate, **25-5**. The chain is next homologated by first displacing the mesylate with the anion from diethyl malonate. Saponification of the ester groups followed by heating the

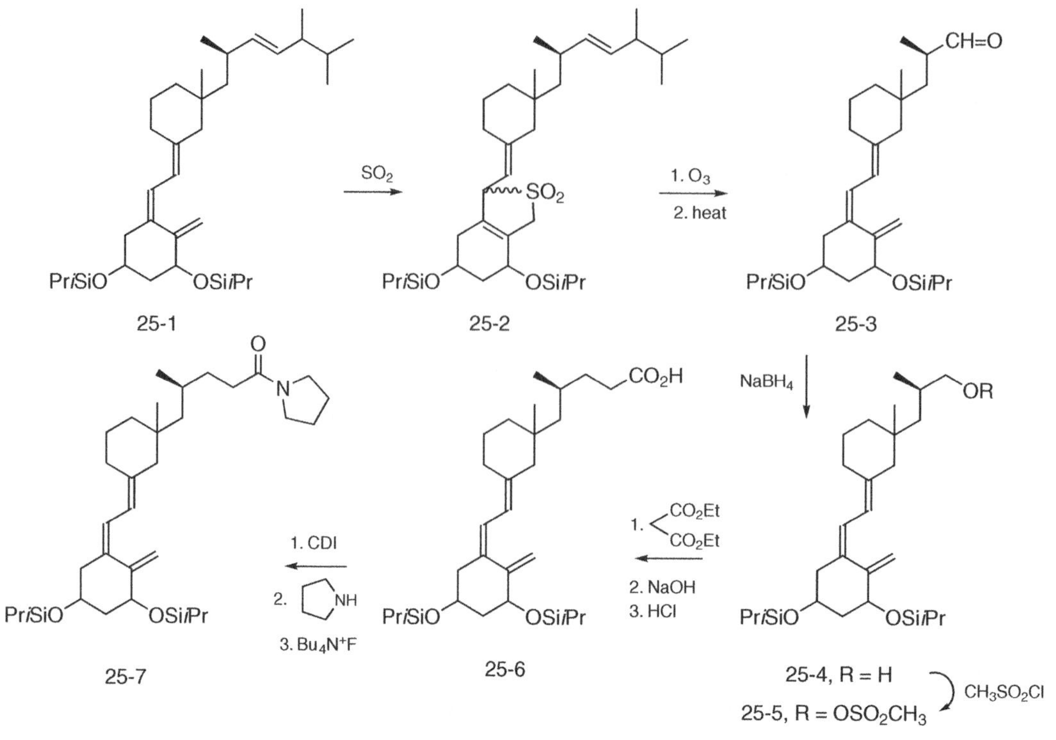

Scheme 8.25

product in acid cause the malonic acid to lose carbon dioxide. There is thus obtained the chain-extended acid **25-6**. The carboxyl group is next converted to the activated imidazolide by reaction with carbonyldiimidazole Condensation of this intermediate with pyrrolidine gives the corresponding amide. Removal of the silyl protection groups with fluoride leads to ecalcidine (**25-7**).

Subject Index

Acetonide, 96
Acetophenide, 96
Acid, cholanic, 8
Acid, cholic, 22
Acid, lithocholic, 8
Acid, Marrianolic, 14
Acid, mevalonic, 20
Aclomethasone, 114
Acne, 133
Adenyl cyclase, 9
Adison's disease, 102
Adrenal glands, hormones from, 16, 102
Adrenalectomy, 16
Advances, pharmacological, 60
Aglycone, 22 129
Aldosterone, 82, 102
Algestone acetonide, 96
Algestone acetophenide, 96
Alkylating agents, selectivity, 45
4-Allylestrone, 38
Altrenogest, 61
American Medical Association, 9
Anabolic activity, 48, 85
Analysis, combustion, 17
    Elemental, 17
Andostane, odor, 80
Androgenic activity, 48, 80, 123
Androgens, circulating, 68
    FDA approved, 85
    Proscribed, 50
Androst-4-ene-17,20-dione,
    from sitosterol, 26
Androstane, 18
Androstane 4,4-dimethyl, 75
Androstane, antineoplastic, 79, 128
Androstedione, from fermentation, 69
Androsterone, 15, 79,
Angesterone, 93
Antagonist, aldosterone, 83
Antagonist, progesterone, 64, 66
Anti-allergic, corticoids, 102, 108, 119, 129
Antiandrogen, 52
Antigonadotrophic, 85, 101
Antihypertensive, 82
Antiinfammatory activity, discovery, 6
Aplastic anemia, anabolic drugs, 67
Approach, conservative, 95
Aromatase, 30, 54
    Inhibitor, 78, 128
Asoprinsil, 65
Asthma, corticoids, 119

Athletes, androgens, 67, 68
6-Aza estrane, 126

Barbier-Wieland, degradation, 14
Bases, optically active, 49
Beef adrenals, source for cortisone, 16
Beta adrenergic blockers, 120
Beta blockers, naming, 9
Betamethasone, 113
Bile salts, from ox bile, 10
Binding, receptor, 9
Bioassay, guide to isolation, 14
Biomimetic, synthesis, 87
Biooxidation, 103
Biopharmacetical properties, 96
Blanc rule, 14
    exception, 13
    in structure studies, 11
Blood, volume, 82
Boar, pheromone, 80
Body builders, androgens, 67, 68, 85
Bolasterone, 79
Burns, anabolic drugs, 67
Butenolide, 130

Cachecia, and cancer, 101
Calcitriol, 133
Calusterone, 79
Cancer, breast, 45, 85, 101
    estrogen dependent, 78
    prostate, 124
Capon units, androgens, 15
Carbocation, 21
Cardiac arrhythmias, 129
    force of contraction, 129
Cell nucleus, 9
Chemotherapy, 45
Chenodeoxycholic acid, 10
Chiral auxillary, 35
Chlormadinone, 92
Cholanic acid, 8
Cholecalciferol, and Vitamin D, 133
Cholestanol, 2
Choleserol, excretion, 15
    1928 structure proposal, 12
    configuration, 12
    exhaustive reduction, 12
    from gallstones, 10
    oxidation in liver, 10
    serum levels, 132
    synthesis inhibitor, 132

Subject Index

Cholic acid, 22, 102
  exhaustive reduction, 12
  from bile salts, 10
Chrysine, from pyrolysis, 11
Clomegestone, 97
Clometherone, 96
Coal tar chemicals, 32
Colestolone, 132
Combustion analysis, 10
Configuration disfavored, 81
  absolute, 4
  hydrindanes; 3
Contraceptive, injectable, 101
  oral, 28, 55, 86, 92
  OTC, 67
  Steroidal, 3
Controlled Substances, list, 50, 68, 71
Coprostane, 12
Corpus luteum, source of progestins, 2, 55, 86
Cortex. Adrenal, 102
Corticoid, 6,9,10, 21-tetrahalo, 117
  6,9,10-trihalo, 116
Corticoids, OTC, 119
  1,4-diene, ubiquity, 109
  9α-F, 11β-OH, ubiquity 109
  topical, 117, 119
Cortisol, 16
Cyclization, squalene, 21
Cycloartenol, 21
Cyclohexanone, latent, 56
Cyclopropyl, from homoallyl, 24
Cyproteron acetate, 93

DEA, 50, 68
Degradation, Barbier-Wieland, 14
Dehydrocholesterol, and Vitamin D, 133
Dehydroepiandrosterone (see DHEA), 5
  Isolated, 15
16-Dehydropregnenolone, 68
3-Deoxy compounds, activity, 60, 63
Deoxycholic acid, 130
Depo-Medrol®, 119
Depo-Provera®, 101
Depots, estrogens, 45
Desogestrel, 64
Dexamethasone, 112
DHEA, 5, 28, 30,
  17-methyl, 73
  stereochemistry, 15
Diascorea villosa, 23
Dicarboxylic acids, cyclization, 11
Diels hydrocarbon, 11
Dienogest, 57
Difluprednate, 114
Digitoxin, 129
Digoxin, 129
5α-Dihydrotestosterone, 50, 68
2,4-Dihydroxyestrone, 38
$6\alpha,17\alpha$-Dimethyltestosterone, 78
2,4-Dinitroestrone, 37

Dioscin, 23
Diosgenin, source of steroids, 23
Diuretic, potassium sparing, 85
  spirolactone, 82
  thiazide, 85
DNA, effect of steroids, 9, 28
Doubly modified, progesterone, 98
Dromostanolone, 70, 74
Drug Enforcement Agency, DEA, 50
Dydrogesterone, 100

Ecalcidine, 134
Edronate drugs, 133
Effector molecules, 9
Electrolites, 9, 82
Endometriosis, 85
Enolization, 3-carbonyl, 71
Enovid®, 67
Enzymes, aromatase, 30
Eplerenone, 85
Equilin, 32
Equilenin, 32, 40
Ergosterol, X-Ray crystallography, 13
Esterase enzymes, 45
Estradiol, 14, 28, 45
  17 esters, 45
  isolation, 14
  nomenclature, 5
Estramustine, 46
Estriol, 14
Estrogen, antagonist, 59
Estrogens, alkylating, 45
  endogenous, 28
  frank, 45
  isolation, 14
  sulfated, 28
Estrone, 14, 28, 30, 34,
  3-methyl ether, 33, 35,36
  commercial synthesis, 34
Estrus inhibition, dogs, 67
Ethisterone, 73, 90
Ethynodrel, 56
Ethynodrone, 56
17α-Ethynyl-estra-3,17β-diol, 45
17α-Ethynylestradiol 3-methyl ether, 56, 67
Etonogestrel, 63
Excretion, cholesterol, 2
Exemestane, 78
Eye drops, 108

Farnesol, 20
Fat, absorption, 2
Feedback inhibitor, cholesterol, 132
Fermentation, 107
  for 11α-hydroxyl, 79, 103
  norgesterel, 61
  sitosterol, 26
  phytosterols, 69
Finasteride, 124
Fluprednisolone, 81
Fluorine, increase in potency, 72, 79

2-Fluorodihydrotestosterone, 73
Fluoromethalone, 113
9α-Fluoroprednisolone, 109
4-Fluoroprogesterone, 89
Fluoxymestrone, 80
Fluticasone, 117
Fluvrestant, 60
Follicle, 55
Follicle Stimulating Hormone, 55
Formestane, 70
Fossils, 2
Foxglove leaves, 129
Fraction, amorphous, 18
FSH, 55
Furan, 122
Furazanoandrostane, 77

Gallstones, source of cholesterol, 10
Genin, suffix, 129
Gestodene, 62
Gestonerone, 58
Glands, endocrine, 16
Glucocorticoid, activity, 102
Glucocorticoids, naming, 3
Glucose, metabolism, 2
Glycosides, 22

Hair growth stimulation, 124
Half-life, orally active drugs, 45
Haloprednone, 115
Handwriting, physicians', 9
Health supplements, 5
Heart, failing, 129
Hecogenin, 126
Heifers, ovulation, 101
Hench, P.S., 102
HMG-CoA reductase, 20
Homoallyl, from cyclopropyl, 24
Hydrindanes, configuration, 3
Hydrocortisone, 16, 17
16α-Hydroxy-17β-estrone, 42
14α-Hydroxyestrone, 3-methyl ether, 43
16α-Hydroxycortisone, 112
2-Hydroxyestrone, 38
Hydroxyprogesterone, 16
17α-Hydroxyprogesterone, 88

Immunosupression, 102
Improvement, additive, 96
Inactivation, bloodstream, 81
    liver, 45, 73
    serum enzymes, 117
    corticoids, 108
Inflammation, 119
    corticoids, 102
    prostaglandins, 102
Inhibitor, receptor binding, 99
INN, 9
Inotropic, 129, 130
Insemination, artificial, 101
Insuflation, corticoids, 119

International Nonproprietary Name (INN) 9
Ion channels, 9
Isoprene, coupling, 2, 20
    Pyrophosphate, 20
Isoxazoloandrostane, 76
IUPC, nomenclature, 2

Johnson, W.S., 32, 87
Julian, Percy, 25

Kendall, E.C, 102
Kinases, 20

Lactation, 86
Lanosterol, 2, 20, 22
Leaving group, 24
Leukotrienes, 9
Lithocholic acid, 8
Lynestrol, 60

Magnetron, NMR, 10
Mare urine, estrogens from, 5, 14, 28
Marker, Russel, 23
Marrianolic acid, 14
Medrogesterone, 95
Medroxyprogesterone, 91
Medrysone, 108
Megesterol, 91
Melengesterol, 99
Menopause, 28, 45, 101
Menstrual cycle, 14, 86
Messengers, secondary, 9
Mestanolone, 73
Mesterolone, 74
Mestranol, 43
Metabolism, carbohydrate, 102
Metabolism, estradiol and estrone, 42
Metabolism, glucose, 2, 16
2-Methoxyestrone, 37
Methyl group, floating, 13
Methyl, replacement by ethyl, 60
6-Methylequilenin, 41
Methylprednisolone, 107
6α-Methylprednisolone, 19, 113
16β-Methylprednisone, 109
Mevalonic acid, 20
Mexrenone, 84
Mibolerone, 53, 67
Mifeprex®, 67
Mifepristone, 64, 67
Mineralocorticoid, activity, 102
Molecule, concept, 10
Molecules, effector, 9
Murray, Herb, introduction of 11 OH, 103
Mustard, nitrogen, 45

Names, colloquial, 2
    generic, 9
    trivial, 6
Naphthalene nucleus, 33
Nitrite, transfer, 30

2-Nitroestrone, 37
Nitrogen mustard, 45
Norbolethone, 51, 67
Norethandrolone, 51, 67
Norethynodrel, 67
Norgestrel, 60
Norgestrone, 57
NSAIDS, mechanism, 102

Odor, urinals, 18, 80
Oil, soy bean, 25
Organs, reproductive, 28
Ovary, source of estrogens, 2
Oviduct, assay for estrogens, 9
Ovulation, inhibition, 55, 86
Ovum, 28, 55, 86
Ox bile, source of bile salts, 10
11-Oxa androstane, 126, 127
17-Oxa androstane, 128
6-Oxa estrane, 125
Oxandrolone, 123
Oxendolone, 52

Paper, flat piece, 36
Peterson, Dury, introduction of 11 OH, 103
Pharmacokinetics, origin, 129
Phenanthrol, 14
Pheromone, boar, 80
Phospholipase, 102
Phosphorillation, kinases, 20
Pill, the, 28, 56
Plan B, 67
Plomestane, 53
Polyisoprenoid, 2
Potency, additive, 105
Potency, enhancement by hydroxyl, 111
  enhancement by methyl, 111
  increase, 58
Prednisolone, 107
Prednisone, 105, 119
Pregnan-2,17-dione, 15
Pregnancy, 2, 86
Pregnanediol, 15
Pregnenolone, 14, 25, 88
Premarin®, 28
Press, androgens, 5
Progesterone, 2, 6, 14, 28, 26, 55,86, 88
  iIsolation, 14
  X-ray crystallography, 14
Prostaglandins, 102
Prostatic hypertrophy, 124
Protein synthesis, 28
Provera®, 91, 100
Provest®, 101
Psoriasis, 133
Pyrazoloandrostane, 76

Quinolone antibacterials, naming, 9

Ratio, sodium:potassium, 85
Reaction, one pot, 99

Reactivity, differing, 26,79, 94, 103
Receptor binding, estrogens, 45
  glucocorticoid site, 64
  progestin site, 64
  glucocorticoid, 102
  nuclear, 9
  transmembrane, 9
Reproductive cycle, 28, 86
Resolution, optical isomers, 49
Retrosteroid, 66, 99
Rhizopus, introduction of 11 OH, 103
Roussel-UCLAF, 67
RU-486, 64, 67

Salmeterol, 120
Saponins, 22, 26
Scission, 17-dihidroxyketone, 79, 81
Secalciferol and Vitamin D, 133
Serendipity, 12, 102
Sesquiterpene, 20
Sex offenders, Depo-Provera®, 101
Shark liver oil, 21
Side effects, corticoids, 102
Sitosterol, 25
Skin growth, regulator, 133
Slow release form, 101
Soy bean oil, 25
Sperm, production, 2
Spinal stenosis, 119
Spiroacetal, 23, 24, 54
Spirobutyrophenone, 84
Spirolactone, 83
Spironolactone, 83, 85
Squalene, 21
Stanazol, 76
Statins, 20
Stenbolone, 71
Steranes, in fossils, 2
Stereochemistry, designation, 4, 6
Stereochemistry, DHEA, 16
Stereochemistry, reactions, 18
Steric hindrance, 18
  Strain, 13
Sterols, 10
Stigmasterol, 22, 25 86
Sulfated estrogens, 28
Superfluous, carbomethoxy, 32
Supplements, health, 5
Swoop, one fell, 24
Synthesis, lengthy, 114, 126, 130
Synthon, 20

Terpene, definition, 20
Terpenes, 2
Testes, source of testosterone, 2, 9, 50
Testolactone, 128
Testosterone, 16, 50, 68
  isolated, 15
  naming, 5
Tetralin nucleus, 33
Thioepoxyandrostane, 76

Thiomestrone, 75
Tibolone, 59
Tour de force, synthetic, 69
Trade names, 9
Transposition, hydroxyl, 12 → 11, 102
Trauma, anabolic drugs, 67
Trenbolone, 54
Triamcinolone, acetonide, 116, 119
Triterpenes, 2, 20

U.S. Pharmacopeia, 9
U.S. Adopted Name (USAN), 9
Upjohn, introduction of 11 OH, 103
USAN, 9
Uterus, 55

Vitamin D, 132

Weak-link analogue, 119
WHO, 9
Wieland, Heinrich, 12
Windaus, Adolf, 12
Workhorse, corticoids, 119
World Health Organization, 9

X-ray crystallography, ergosterol, 13

Yam, Mexican, 23, 57

Reactions Index

Note: Entries in this index by and large reflect typical examples of cited reactions. This list, however, does not cite every occurrence in the text of any of the transforms listed below.

Acetal, 17–21, formation, 114
  bismethylenedioxy, formation, 10
  formation, 54, 58, 62, 64, 65, 90, 94–97, 108, 110, 112, 117, 126
    acetophenone, 96
    tetrahydropyran, 90
  interchange, 69
Acylation, 31, 60, 91 bromine to olefin, 57
  carboxyl, acetylene, 83
  chlorine, olefin 110, 116, 117
  hydrogen chloride, eneone, 114
  hypobromous acid, to olefin, 31, 54, 80, 102, 108–111, 113, 115, 116
  hypochlorous acid, to olefin, 52
  lithio reagent to epoxide, 45, 54
  perbromomethyl, enol ether, 93
  sulfur didoxide to 1,3-diene, 134
Alkyl-1,3-cyclopentanone, condensation, 34, 35, 125, 126
Alkylation, base mediated, 40, 74, 81, 90
  base, iodomethane, 32, 33
  carbanion, C17, 95, 100
  acetophenone, 38, 59
Ambident anion, 34
Amidation, imidazolide, 124
Aromatization, 1,4-dien-3-one, 29, 57, 58
  bond migration, 37

Backside attack, 16, 18, 34, 43, 61, 56, 61, 70, 78, 81, 92, 108, 109
Base catalyzed, extrusion of 10 methyl to phenol, 29
  fusion, 16-hydroxyestradiol, 14, 58
Birch reduction, see Reduction, Birch,
Bismethylenedioxy acetal, 106
Bromination, 1,4-ene-3-one, 76
Bromination, C$_2$, 75, 123
  C$_2$,C$_4$, 28, 57
  bromine, 99, 103, 105, 108, 112, 115, 127
  furan, 130
  bromohydantoin, 100
  allylic N-bromosuccinimide, 132
Bromonium ion, 54, 99, 109
Butenolide formation, 131
tert-Butyldimethylsilyl ether TBDMS, 43
Butyrolactone spiro, formation, 83

Carbethoxy, from nitrile, 84
Carbethoxylation, Wittig, 131
Carbonylation, palladium mediated, 39, 44

Carboxylation, acetylene, 83
Chiral auxillary, 35
Chlorination, chlorosuccinimide, 92
  N-chloroacetamide, 110
  olefin, chlorine, 96
  perchloric acid, 72
  sulfuryl chloride, 72
Chloroketones from acid chloride and diazomethane, 17
Chloronium ion, 52, 96
Colldine, see dehydrobromination,
Condensation aldol, acetaldehyde, 81
  aldol, oxalate 103
  Diels Alder, 35
  Favorskii, 33
  Knovenagel, 51
  Stobbe, 32
  Wittig, 36, 131
Conjugate addition, 49, 53, 58, 59, 94, 100
  (1,6), 79
  (1,6) cyanide, 84
  thioacetate, 84
  to 9(11)-5(10)-epoxide, 64–66
Cyanohydrin, formation, 61, 83
Cyclization, 1,6-ene, titanium chloride, 127
  base catalyzed, 32, 37
  dicarboxylic acid, 11
  diester, Dieckman, 33, 34, 36
  diol, 11-oxasteroid, 127
  enamine mediated, 49, 50
  free radical, 36
  Friedel-Crafts, 32
    aryloxypropionic acid, 125
  internal Diels Alder, 69, 128
Cyclodehydration, diacid, 14
Cyclopropanone, formation, transient, 103
Cyclopropyl formation, chloromethyl, 93
  formation pyrrazole, 92
  from, homoallyl, 24, 31
  opening, acid, 92

Debromination, 57
  iodide, 99
Decarboxylation, ketoester, 32, 34
  malonic acid, 134
Degradation, Barbier-Wieland, 14
  oxidative' 13
Dehalogenation, pyridine, 93

Steroid Chemistry at a Glance   Daniel Lednicer
© 2011 John Wiley & Sons, Ltd

Dehydration, 107
  11-hydoxyl, 63
  acid catalyzed, 15, 18, 54, 93
  boron trifluoride, 123
  phosphorus oxychloride, pyridine, 109
  solvolysis, 24
  thionyl chloride, 116
  β-hydroxyketone, 70, 78, 80, 90, 91, 99
Dehydrobromination, base, 29, 71, 74, 80, 112, 115, 127
  collidine, 28, 29, 41, 57, 93, 97, 100, 103, 108, 123, 132
Dehydrochlorination, acetate, 96
Dehydrogenation, chloranil, 52, 53, 58, 79, 95
  DDQ, 41, 42, 54, 69, 108, 110, 114, 124, 128
  selenium, 11
  selenium dioxide, 73, 92, 104, 108, 111, 112, 113, 116
Deoxygenation, dithioacetal, nickel, 60, 81, 93
  sodium, 63
  phenol phosphite, 40
Diaxial opening, oxirane, 78, 81, 90
Diazoketone, formation, 17
Diazomethane, addition to olefin, 73, 92
  reaction with acid halide, 17
Diborane, oxygenation, 42
Diels Alder, see Condensation,
Dihydroxylation, osmium teroxide, 42, 70, 91, 94, 104, 112, 114, 116
  permanganate, 96
Dipolar addition, 1,3, diazomethane, 74
Displacement, halogen by acetate, 16, 17, 105
  hydroxyl, phosphorus pentachloride, 16
  iodine, hydroxyl group, 127
  mesylate, chlorine, 117
  mesylate, iodide, 113
  mesylate, malonate, 134
  tosylate, acetate, 81

Electrocyclization, 36, 69
  mesylate to olefin, 41
  acetate, heat mediated, 43
  mesylate, 115
  β-hyroxyl, base catalyzed, 24
Enamine formation, 26, 50, 53, 63, 68, 79, 80, 89,
Enol acyl, formation, 42, 78, 94, 104, 111, 115,
  ether, formation, orthoformate, 110, 127,
  iodide, 44,
  lactone, 49,
Epoxidation, basic hydrogen peroxide, 57, 85, 98,
  from 1,2-bromohydrin, 54, 81
  from chlorohydrin, 127,
  from trimethylsulfonium ylid, 66,
  m-chloroperbenzoic acid, 43, 78, 91, 94, 122
  monoperphthalic acid, 54,
  peracetic acid, 41, 76, 99,
  peracid, 94,
  perbenzoic acid, 42, 92, 104, 106, 107,
  perfloroacetone peroxide, 64,
  perphthalic acid, 87, 90
  trifluoroperacetic acid, 122
Epoxide opening, azide, 118
  hydrogen bromide, 58
  hydrogen cyanide, 57

hydrogen chloride, 92
hydrogen fluoride, 51, 54, 81, 99, 109, 110, 111, 113, 115, 116
  thiocyanate, 76
  formation, bromohydrin, 81, 109, 110, 111, 113, 115, 116
  rearrangement, 122
Ethynylation, acetylene, 43, 59, 60, 73, 125, 127

Favorskii reaction, 33
Fluorination, fluorine perchlorate, 111, 115
Fluorination, β-formylketone, fluorine perchlorate, 72
Fluorohydrin, formation, 55, 83, 109, 110, 111, 113, 115, 116
Fluoromethyl thioester, formation, 119
Formylation, DMF, phosphorus oxychloride, 117
  ethyl formate, 32, 70, 74
  Vilsmeyer, 91
Free radical, electrocyclization, 36
Furan, formation, 130
Furazole, from oximinoketone and hydroxylanine, 77

Grignard, addition, aldehyde, 133
  9(11)-5(10)-epoxide, 64–66
  11-ketone, 62
  carboxylic ester, 15
  enol lactone, 49
  epoxide, 87, 90, 100, 106
  ketone, 24, 34, 35, 43, 51, 61, 73, 74, 79, 80
  conjugate, 112
  reagent, from 4-halobenzenes, 64, 65, 66
  vinylmagnesium, 34, 122, 124–126
  2-propenyl, 133

Hemiacetal, formation, 18
Homologation, ester, 34
Hydrindanone, formation, 34, 36
Hydroboration, 62
Hydrogenation catalytic, olefin, see Reduction,
Hydrolyisis, cyanohydrin, 61, 64, 130
Hydrolysis, acetal, 37, 41, 54, 62, 65, 66, 67, 83, 87, 94, 96, 107–113, 116, 126
  acetal bismethylenedioxy, 107
  acetal, tetrahydropyran, 90
  enamine, 50, 68, 80, 89
  enol ether, 48, 51, 56, 57, 59, 63, 68, 73, 91, 92, 127
  ester, 45, 88
  imine, 84, 88
  nitrile, 83
  oxime, 61
  sugar, 23, 130
Hydroxylamine, oxime formation, 32
Hypobromous acid see Addition,

Imminium salt, formation, 44
Inversion, B/C ring fusion, 66, 100
  configuration, 16
Isoxazole, from β-formylketone and hydroxylamine, 76
Isoxazoline, formation, 32

Ketal, *see* acetal

Lactam, formation, 124, 129
Lactone formation, 31, 128

Mannich reaction, 71
Mesylate, formation, 62, 113, 115, 117
Methylation, diazomethane, 108
Methynylation, enol ether, Vilsmeyer, 79
Migration, olefin $C_4 \rightarrow C_5$, 106
    $C_{5(9)} \rightarrow C_4$, 51, 58
    $C_5 \rightarrow C_4$, 28, 29
    $C_8 \rightarrow C_{9(11)}$, 62
    $C_7 \rightarrow C_8$, 132
Mitsonobu acetoxylation, 131
  halogenation, 36
Mixed anhydride, diphenylphosphate, 119

Nitration, 38
Nitrosation, ketone, butyl nitrite, 77
Nitrosobromination 3-acetoxy-5-ene, 31

Olefination, see Condensation, Wittig,
Oxazolidine, spiro, formation, 83
Oxazoline, formation, 118
Oxidation, basic conditions, 37
  Bayer-Viiliger, 128
  chromium trichloride, 17, 18, 24, 25, 31, 36, 42, 53, 79, 84, 97, 102, 103, 107, 126, 132
  diborane adduct, hydrogen peroxide, 63
  enzymatic, 30
  epoxide formation, see epoxidation,
  Fremy salt, 38
  Haloform-like, 123
  Jones reagent, 43, 52, 58, 63, 96, 104
  lead tetraacetate, 31, 52, 128
  manganese dioxide, 69, 131
  N-bromosuccinimide, 108
  N-methylmorpholine oxide, 104
  Oppenauer, 16, 17, 26, 28, 29, 51, 56, 57, 59, 73, 78, 83, 87, 94, 95, 99, 100
  osmium tetroxide, 70
  periodate/permanganate, 123
  photo, 26, 131
    α to carbonyl, 131
  progesterone, fermentation, 103
  pyridinium chlorochromate, 40, 50, 52, 62, 63, 90
  singlet oxygen, 69
  sulfide, 60
  Swern, 35, 130, 131
  unspecified, 11, 12, 14, 15, 16, 18, 21, 38, 53, 61, 68, 84, 98, 102, 108, 112, 128
Oxime, formation, 32, 68, 129
Oxygenation, diborane, 42
  lead tetraacetate, 123
Ozonization, 26, 126, 134
Phosgene, urethane formation, 45
Photo-isomerization, 66, 100
Photo-oxygenation, 26
Prins reaction, 131
Protecting groups, juggling, 69
Pyridone, formation, 124

Pyrolysis, 11
  extrusion of 10 methyl to phenol., 29
Pyrrazole formation, diazomethane, 74, 92
  from β-formylketone and hydrazine, 76
Pyrrolidine, formation, 129

Radical di-anion, 48
Rearrangement, Beckman, 68, 129
  Claisen, 38
  dienone-phenol, 29
  epoxyketone, 122
  Favorsky, 103
  homoally-cyclopropyl, 24, 31
  nitrile, 84
  photolytic 5-bromo 6-nitrite to 19-oxime, 31
  photolytic 6-nitrite to 19-oxime, 30
  thioformamide $O \rightarrow S$, 119
Reduction
  Birch, 34, 35, 48–60, 63, 64, 67, 95
    phenol phosphite, 40
    quinoline, 126
  bromine, zinc, 102
  butenolide, diisobutylaluminum hydride, 130
  carbonyl, borohydride, 34, 37, 113, 125, 126, 134
    lithium alumnum hydride, 40, 42, 43, 51, 70, 80, 81, 107, 126
    sodium borohydride, 40
    tert-butoxyaluminum hydride, 122
    unspecified, 12, 18
  catalytic, 16, 28, 34, 35 49, 50, 52, 70, 71, 81, 72, 91, 93, 94, 108, 125, 126, 129
    acetylene, 83
    azide, 118
  disiobutylaluminum hydride, 84, 130
  epoxide, lithium aluminum hydride, 43, 133
  iodide, bisulfite, 99
    zinc, 112, 113
  ketone, tri-tbutoxyhydride, 122
    yeast, 16
    amalgamated zinc, 15
  lactol, borohydride, 123
  nitrile, diisobutylaluminum hydride, 84
  oxime, lithium aluminum hydride, 44
  tin, 37
  unspecified, 12, 15, 18, 32
  zinc, halogen, 58
    hydroxyl, 132
Resolution, enantiomers, 49
Reversal sulfur dioxide addition product, 134
Ring opening, diaxial, 31, 41, 54, 58, 64, 76, 78, 108, 122
  photolytic, 133

Saponification, 17, 31, 34, 49, 52, 54, 60, 64, 81, 89, 96, 97, 99, 100, 105, 108, 117, 119, 126, 132, 134
Scision, 1,2-diol, periodate, 126
  17,21-dihydroxy-20-ene, bismuthate, 80
  CO bond, boron trifluoride, 31
  tert-butyldimethylsilyl, 43
  cyclohexanone ring, 13, 124, 132
  hydroxyketone, periodate, 18
  methyl ether, 53

Scision, 1,2-diol, periodate (*Continued*)
   oxalate, 103
   side chain, periodate, 118
   silyl ether, 36, 59, 124, 134
   spiroacetal, acetic anhydride, 24
   trimethylsilyl cyanide, 66
Sequence, now familiar, 115
i-Steroid rearrangement, see rearrangement, homoallyl-cyclopropyl

Thioacetal, formation, 60, 63, 81, 93
Thioepoxide, from 2-hydroxyisothiocyanate, 76
Thioester formation, 117
Thiomethylation, 75
Tosylation, 25, 31, 80, 81
Transposition, oxygen, $C_{12} \rightarrow C_{11}$, 102
Triflate, displacement, 39
Trimethylsilyloxy nitrile, formation, 64
Trimethylsulfonium ylide addition to ketone C=O bond, 65, 66, 84

Vilsmeyer reaction, 78, 91

Printed and bound by CPI Group (UK) Ltd, Croydon, CR0 4YY
09/06/2025
14685659-0003